Professional Development Manual

STATISTICS:
A Key to
Better Mathematics

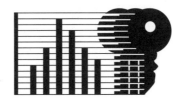

**The University of North Carolina
Mathematics and Science Education Network**

Dale Seymour Publications®

Project Editor: Joan Gideon
Production Coordinator: Claire Flaherty
Design Manager: Jeff Kelly
Text and Cover Design: Don Taka

Published by Dale Seymour Publications®, an imprint of
the Alternative Publishing Group of Addison-Wesley Publishing Company.

The Teach-Stat materials were prepared with the support of National
Science Foundation Grant No. TPE-9153779. Any opinions, findings,
conclusions, or recommendations expressed in this publication are
those of the authors and do not necessarily represent the views of the
National Science Foundation. These materials shall be subject to a royalty-free,
irrevocable, worldwide, nonexclusive license in the United States Government
to reproduce, perform, translate, and otherwise use and to authorize others
to use such materials for Government purposes.

 This project was supported, in part, by the
National Science Foundation
Opinions expressed are those of the authors
and not necessarily those of the Foundation

Additional funding was provided by the North Carolina Department of
Public Instruction and the North Carolina Statewide Systemic Initiative.

Copyright © 1997 The University of North Carolina. All rights reserved.
Printed in the United States of America.

Limited reproduction permission. The publisher grants permission to individuals
who have purchased this book to reproduce the blackline masters as needed for
use with teachers. Reproduction for an entire school or school district or for
commercial use is prohibited.

Many of the designations used by manufacturers and sellers to distinguish their
products are claimed as trademarks. Where those designations appear in this book
and Addison-Wesley was aware of a trademark claim, the designations have been
printed with initial capital letters.

Excerpts from **Rules of Thumb II** Copyright © 1987 by Tom Parker. Reprinted by
permission of Houghton Mifflin Company. All rights reserved.

Order Number DS21407
ISBN 0-86651-950-5

2 3 4 5 6 7 8 9 10-ML-99 98 97 96

This book is printed
on recycled paper.

Co-Principal Investigators:
Statewide Implementation
Susan N. Friel, UNC Mathematics and Science
  Education Network
Jeane M. Joyner, North Carolina Department of
  Public Instruction

Editors
Susan N. Friel, UNC Mathematics and Science
  Education Network
Jeane M. Joyner, North Carolina Department of
  Public Instruction

Co-Principal Investigators: Local Implementation
George W. Bright, University of North Carolina
  at Greensboro
Theresa E. Early, Appalachian State University
Dargan Frierson, Jr., University of North Carolina
  at Wilmington
Jane M. Gleason, North Carolina State University
M. Gail Jones, University of North Carolina at Chapel Hill
Robert N. Joyner, East Carolina University
Nicholas J. Norgaard, Western Carolina University
Mary Kim Prichard, University of North Carolina
  at Charlotte
Clifford W. Tremblay, Pembroke State University

Evaluation Team
Sarah B. Berenson, North Carolina State University
George W. Bright, University of North Carolina
  at Greensboro
Diane L. Frost, Curriculum and Instruction, Asheboro
  City Schools, Asheboro, North Carolina
Lynda Stone, University of North Carolina at Chapel Hill

Curriculum Consultants
Gwendolyn V. Clay, Meredith College
Gary D. Kader, Appalachian State University
Karan B. Smith, University of North Carolina
  at Wilmington

Editorial and Production Consultants
Stephanie J. Botsford, UNC Mathematics and Science
  Education Network
Stacy L. Otto, UNC Mathematics and Science
  Education Network
Elizabeth M. Vesilind, University of North Carolina
  at Chapel Hill

Administrative Support
Sherry B. Coble, Center for Mathematics and Science
  Education, University of North Carolina at Chapel Hill
Kathleen J. Gillespie, UNC Mathematics and Science
  Education Network
Jane W. Mitchell, UNC Mathematics and Science
  Education Network
Maurice A. Wingfield, Jr., Mathematics and Science
  Education Center, Fayetteville State University

Graduate Assistants
Anita H. Bowman, University of North Carolina
  at Greensboro
Lynne C. Gregorio, North Carolina State University
Sarah J. Pulley, East Carolina University
Julie C. Shouse, University of North Carolina
  at Wilmington
Kathryn L. Speckman, University of North Carolina
  at Greensboro
Kimberly I. Steere, University of North Carolina
  at Chapel Hill

Teach-Stat Advisory Board
Gloria B. Barrett, North Carolina School for Science
  and Mathematics
Randy L. Davis, Glaxo, Inc.
Denis T. DuBay, North Carolina Science and Mathematics
  Alliance
Diane L. Frost, Curriculum and Instruction, Asheboro
  City Schools, Asheboro, North Carolina
Manuel Keepler, North Carolina Central University,
  Department of Mathematics and Computer Science
L. Mike Perry, Mathematics Department, Appalachian
  State University
Sherron M. Pfeiffer, Equals, Hendersonville,
  North Carolina
Jerome Sacks, National Institution of Statistical Science
Michael J. Symons, Department of Biostatistics,
  University of North Carolina at Chapel Hill
John L. Wasik, Department of Statistics, North Carolina
  State University
Johnny F. Warrick, Robinson Elementary School,
  Gastonia, North Carolina

UNC Mathematics and Science Education Network
Center Directors
Sarah B. Berenson, North Carolina State University
J. Ralph DeVane, Western Carolina University
Steven E. Dyche, Appalachian University
Leo Edwards, Jr., Fayetteville State University
Vallie W. Guthrie, North Carolina A & T State University
Katherine W. Hodgin, East Carolina University
Russell J. Rowlett, University of North Carolina
  at Chapel Hill
Josephine D. Wallace, University of North Carolina
  at Charlotte
Charles R. Ward, University of North Carolina
  at Wilmington

APPENDIX: PARTICIPANT TEACHERS

Year One Teachers

Irene Baldwin
T. J. Blake
Carol Blankenhorn
Anessa Burgman
SuAnn Burton
Cynthia Collins
Elaine Corbett
Lisa Ann Crocker
Mary Lee Danielson
Rita Davis
Rebecca Deal
Ann Diedrick
Barbara Dishman
Doris Farmer
Marie Flynt
Mary Anne Simon Frost
Robin Frost
Sue Ellen Goldstein
Barbara Ann Gustafson
Eileen Hartwell
Donna Hash
Angela Heustess
Deborah Hill
Mary Ann Hopkins
Carla Jacobs
Donna James
Donna Jenkins
Denise Jewell
Sharon Frances Leak Jones
Rosemary Klein
Cathy Jo Korenek
Gail Lane
Jacqueline Lauve
Linda Law
Sharon Leonard
Charisma Lindberg
Amy Loy
Ron Luciano
Rhonda McCurry
Kay Moore
Phyllis Moore
Teresa Pace
Wendy Rich
Paula Rigsbee
Paula Segers
Linda South
Linda Stroupe
Sandra Styron
Frances Toledano
Cheryl Varner
Jane Wagner
Harriet Weinstock
Anne White
Deborah Whorley
Julia Wilkie
Sheila Wilkins
Vicki Yoder

Year Two Teachers

Pat Adkisson
Sandra Albarty
Etta Alston
Donna Anderson
Mary Backes
Brenda Bailey
Kathy Baily
Debra Baize
Lisa Baker
Rosemary Barker
Jennie Beedle
Kim Beeson
Cynthia Bell
Brenda Berry
Suzanne Billips
Sandra Billups
Martha Blount
Charles Brantley
Kimberly Briles
Hilda Bukowski
Barbara Burns
Gregg Byrum
Hattie Campbell
Angela Cannon
Ann Cannon
Mariea Carey
Bennie Carpenter
Alissa Carver
Evelyn Case
Lisa Celotto
Betsy Church
Leticia Clark
Sue Clayton
Nancy Cooke
Marsha Corbett
William Cornett
Beth Cornwell
Bettie Council
Susan Creasy
Janice Croasmun
Pamela Crowhurst
Becky Crump
Dinah Cuthbertson
Carol Davidson
Judy Davidson
Debra Davis
Sylvia Davis
Ann Dawson
Patricia Denson
Susan Douglas
Terry Dryman
Donna Dysert
Roberta Ebron
Donna Edmiston
Ruth Ann Edmunds
Barbara Elder
Angela Farrar
Brookie Ferguson
Diane Flagler
Jennifer Flesca
Bonnie Flynn
Phyllis Forbes
Judy Foster
Dorothy Freeman
Jennifer Freeman
Denise Fuller
Marcia Gallant
Debra Gandy
Barbara Gaw
Angela Gaylor
Nancy Glenham
Joan Goff
Clarke Goodman
Phillip Gordon
Beth Graves
Maria Grigg
Sarah Gustafson
Lisa Haley
Greg Hathcock
Cheryl Hawkins
Katherine Hayes
Catherine Heglar
Montrose Helms
Susan Henderson
Lisa Henningan
Annie Hicks
Helen Hollis
Patty Honeycutt
Sharyn Hoover
Gail Horne
Janice House
Amanda Hykin
Barbara Jackson

Jo Johnson
Sandy Johnson
Julie Jones
Lois Jones
Sandra Jones
Carolyn Joplin
Saralyn Kader
Julie Kappers
Margaret Kelder
Virginia Kelly
Frances Kemp
Betsy Kinlaw
Becky Kirkendall
Irene Koonce
Flora Annette Kremer
Jeanne Lamb
Vivian Lamm
Benita Lawrence
Karen Lawrence
Addie Leggett
Mary Lipscomb
Diane Livingston
Leah Lowry
Margaret Lundy
Lorraine Malphurs
Donald Mapson
Camille Marlowe
Grace Martin
Nancy Martin
Cathi McClain
Margaret Anne McColl
Debbie McCord
Virginia McElroy
Diane McFerrin
Donald Rose McGhee
Kitty McGimsey
Dorothy Gray McKoy
Tina McSwain
Clyde Melton
Kathy Mizerak
Jackie Monroe
Barbara Moore
Louise Moore
Linda Moose
Aundra Moretz
Mary Morrow
Pat Mullens
Michael Mulligan
Karen Myers
Cheryl Newby
Debra O'Neal
Brenda Oxford

Shelia Page
Karen Parham
Cynthia Parker
Barbara Paul
Deborah Pevatte
Janice Phillips
Keyna Pittman
Molly Pope
Betty Powell
Jane Powell
Kim Berry Price
Donese Pulley
Virginia Reed
Cheryl Ricks
Charlotte Riddick
Carla Rierson
Karen Rodenhizer
Deborah Ross
Lynne S. Rozier
Barbara Seaforth
Ramona Sethill
Jacqueline Setliff
Barbara Shafer
Jean Slate
Cynthia Smith
Jean M. Smith
Pamela Smith
Susan Parker Smith
Terry Smith
Stephen Sorrell
Joan Stafford
Brenda Stanley
Polly Stewart
Wanda Sutton
Judy Swain
Betty Sykes
Katherine Taylor
Janet Thompson
Kimberly Thompson
C. Henry Thorne
Debbie Thorsen
Jennifer Tilley
Allene Trachte
Kathy Tramble
Carolyn Tryon
Gail Uldricks
Mary Clay Vick
Carol Wainwright
Linda Ward
James R. Watson
Priscilla Waycaster
Doris Weaver

Glenda Weaver
Sarah Wells
Patsy Whitby
Susan Whitehurst
Joseph N. Whitley, Jr.
Harriette Whitlow
Peggy Lee Wilder
Hollis Williams
Judy Williams
Wanda Williams
Myrtle Winstead
Jacequelin Wiseman
Brenda Woodruff
Kimberly Woodruff
Linda Wright
June Zurface

Contents

Introduction to the Professional Development Manual xiii

Teach-Stat Workshop: Three-Week Syllabus xxi

Investigation 1   About Us 1

    Transparency 1.a   Process of Statistical Investigation
    Transparency 1.b   The Concept Map of the Process of Statistical Investigation
    Handout 1.1   The Statistical Investigation Process

Investigation 2   Sorting People: Who Fits My Rule? 15

Investigation 3   Yekttis 19

    Handout 3.1   Using Venn Diagrams: Sorting and Classifying Yekttis

Investigation 4   Sorting Things! 27

Investigation 5   Restructuring Mathematics 31

    Transparency 5.a   Guidelines for the Modern Teacher
    Transparency 5.b   NCTM *Professional Standards for Teaching Mathematics*
    Transparency 5.c   Assumptions About Teaching Mathematics
    Transparency 5.d   Standards for Teaching Mathematics
    Handout 5.1   Using the *Professional Standards for Teaching Mathematics*

Investigation 6   Shape of the Data: Using Line Plots 39

    Transparency 6.a   Overall Shape of the Data
    Transparency 6.b   Distribution A
    Transparency 6.c   Distribution B
    Handout 6.1   Distribution A, Distribution B

Investigation 7   Shape of the Data: Line Plots to Bar Graphs 49

    Transparency 7.a   Letters in Names Line Plot and Frequency Table
    Transparency 7.b   Letters in Names Line Plot and Bar Graph
    Transparency 7.c   Three Components to Graph Comprehension
    Handout 7.1   Line Plots to Bar Graphs Problem Sheet
    Handout 7.2   Three Components to Graph Comprehension

Investigation 8   **Giant Steps, Baby Steps** 59

    Transparency 8.a   Giant Steps, Baby Steps: Summary
    Transparency 8.b   Rules of Thumb

Investigation 9   **What Is the Typical Foot Length of Our Group?** 67

    Transparency 9.a   Typical Foot Length: Summary

Investigation 10   **Children and Measurement: A Minilecture** 75

    Transparencies 10.a through 10.m

Investigation 11   **Accuracy in Measurement** 97

    Handout 11.1   Accuracy in Measurement Problem Sheet

Investigation 12   **How Close Can You Get to a Pigeon?** 103

Investigation 13   **Family Size** 107

    Transparency 13.a   Family Size Bar Chart: Ungrouped Data (Unordered)
    Transparency 13.b   Family Size Bar Chart: Ungrouped Data (Ordered)
    Transparency 13.c   Family Size Bar Graph: Grouped Data
    Handout 13.1   Family Size

Investigation 14   **Median: More Than Just the Middle of the Data** 117

    Transparency 14.a   Roofs 1
    Transparency 14.b   Roofs 2
    Handout 14.1   Roofs 1
    Handout 14.2   Roofs 2

Investigation 15   **Types of Data: A Minilecture** 129

    Transparencies 15.a through 15.f

Investigation 16   **About Us Revisited** 139

    Handout 16.1   Categorical Data, Mode, and Range Problem Sheet

Investigation 17   **Shape of the Data: Using Stem Plots** 145

    Transparency 17.a   Shapes of Stem-and-Leaf Plots
    Handout 17.1   Distribution A, Distribution B

Investigation 18	**Shape of the Data: Stem Plots to Histograms**	**155**
Transparency 18.a	Minutes to Travel to School: Stem-and-Leaf Plot	
Transparency 18.b	Minutes to Travel to School: Frequency Table	
Transparency 18.c	Minutes to Travel to School: Histogram (10-minute intervals)	
Transparency 18.d	Minutes to Travel to School: Histogram (5-minute intervals)	
Transparency 18.e	Travel Time to School	
Transparency 18.f	Allowances of 60 Students: Histogram ($.50 intervals)	
Transparency 18.g	Allowances of 60 Students: Histogram ($1.50 intervals)	
Transparency 18.h	Allowances of 60 Students: Histogram ($1.00 intervals)	
Handout 18.1	Stem Plot to Histogram Problem Sheet	

Investigation 19	**Curriculum Overview**	**171**

Investigation 20	**Shape of the Data: Problem Solving**	**173**
Handout 20.1	Shape of the Data Problem Sheet	

Investigation 21	**Graphing: A Minilecture**	**183**

Transparencies 21.a through 21q

Investigation 22	**Raisins Revisited**	**211**
Handout 22.1	Raisins Problem Sheet	

Investigation 23	**How Do We Grow?**	**217**
Transparency 23.a	Massachusetts and Georgia Classes Bar Graphs	
Transparency 23.b	Raisins Histograms (Uncorrected Scales)	
Transparency 23.c	Raisins Histograms (Corrected Scales)	
Transparency 23.d	Massachusetts Class Bar Graph and Data Values	
Transparency 23.e	Georgia Class Bar Graph and Data Values	
Transparency 23.f	Massachusetts Class Bar Graph, Data Values, and Box Plot	
Transparency 23.g	Georgia Class Bar Graph, Data Values, and Box Plot	
Transparency 23.h	Massachusetts and Georgia Classes Box Plots	
Handout 23.1	Massachusetts and Georgia Classes Bar Graphs	
Handout 23.2	Height Data, Sheet 1	
Handout 23.3	Height Data, Sheet 2	

Investigation 24	**Raisins Revisited, Revisited**	**235**
Transparency 24.a	Raisin Data Line Plots	
Transparency 24.b	Raisin Data Box Plots	
Handout 24.1	Two Brands Data	

Investigation 25	**Graphing Data Using Computers**	**241**
Handout 25.1	Raisins in a Half-Ounce Box	
Handout 25.2	Cereal Data	

Investigation 26   **Family Size Revisited** **245**

  Transparency 26.a   The Mean as a Balance Point: Data Set 1
  Transparency 26.b   The Mean as a Balance Point: Data Set 2
  Handout 26.1   The Mean as a Balance Point
  Handout 26.2   Comparing Balance Points
  Handout 26.3   More Work with the Mean

Investigation 27   **Cats** **257**

  Handout 27.1   Cat Data Collection Sheet

Investigation 28   **Technology** **263**

  Handout 28.1   Exploring Cats: Entering Records
  Handout 28.2   Exploring Cats: Finding Records
  Handout 28.3   Exploring Cats: Sorting Records

Investigation 29   **Building the "Rule" for Finding the Mean** **269**

  Handout 29.1   Survey Results
  Handout 29.2   Mean Problem Sheet

Investigation 30   **Means in the News** **275**

  Handout 30.1   Real Means

Investigation 31   **Comparing Sets of Cereal Data** **279**

  Transparency 31.a   Cereal Data Line Plots
  Transparency 31.b   Cereal Data Box Plots
  Transparency 31.c   Cereal Data Line Plots and Box Plots
  Handout 31.1   Sugar Content of Common Foods
  Handout 31.2   Cereals on the Bottom Shelf
  Handout 31.3   Cereals on the Middle Shelf
  Handout 31.4   Cereals on the Top Shelf

Investigation 32   **Students' Understanding of the Mean: A Minilecture** **289**

  Transparencies 32.a through 32.h

Investigation 33   **Classroom Assessment of Mathematics: A Minilecture** **301**

  Transparencies 33.a through 33.n

Investigation 34   **Linking Probability and Statistics: A Minilecture** **321**

  Transparencies 34a, 34.b

Investigation 35	**True-False Test**	327
Investigation 36	**Removing Markers from a Number Line**	331
Handout 36.1	Removing Markers Game Sheet	
Investigation 37	**What Are the Odds?**	335
Handout 37.1	Rolls of a Pair of Dice	
Investigation 38	**Fair Games**	339
Handout 38.1	Fair Games Data Sheet	
Investigation 39	**What's in the Bag?**	343
Investigation 40	**Choosing Samples**	347
Handout 40.1	100 Cats Database	
Handout 40.2	Total Cat Data Summary	
Handout 40.3	Female Cat Data Summary	
Handout 40.4	Male Cat Data Summary	
Investigation 41	**How Tall Are You?**	357
Investigation 42	**Are You a Square?**	363
Transparency 42.a	Leonardo's Square Man	
Investigation 43	**From Footprint to Stature**	367
Investigation 44	**Cats Revisited**	371

Introduction to the Professional Development Manual

Numbers, data, and information surround us. Tables, graphs, charts, and statistics abound in newspapers, magazines, advertisements, and on radio and television. We are sometimes confronted with reports involving false uses of data. Decisions about our home, health, finances, and use of our time are all affected by data. To operate effectively in our complex world, we must be able to interpret graphs, charts, and tables, as well as understand—not merely compute—basic descriptive measures.

In short, to be successful in this society we need "data sense." Having data sense means being comfortable with posing questions, collecting and analyzing data, and interpreting the results in ways that respond to the questions we originally asked. It implies comfort and competence in reading and evaluating reports that are based on statistics. And it is more than just understanding graphs and statistics; it involves asking questions about the *process* of statistical investigation.

The Process of Statistical Investigation

Statistics is the science that uses data to answer questions and enables us to make wise decisions in the face of uncertainty. In a sense, statistics is numerical descriptions of quantitative aspects of "things"—people, objects, locations—and may take the form of counts or measurements. Statistics, however, also refers to a "subject"—methods for obtaining and analyzing data in order to make decisions that may focus on questions of practical action or scientific research.

What makes a problem a statistical investigation is the way the question is asked, the role and nature of the data, the way in which the data are examined, and the types of interpretations made from the examination.

A statistical investigation typically contains the following four components—the four stages of the PCAI model (Graham 1987)—in some order:

- **Pose the question** Identify a specific question to explore, and decide what data to collect to address the question.

- **Collect the data** Decide how to collect the data, and collect the data.

- **Analyze the data** Organize, summarize, describe, and display the data; and look for patterns in the variation of the data.

- **Interpret the results** Use the results from the analyses to make decisions about the original question.

We can visualize the process of statistical investigation in the following way:

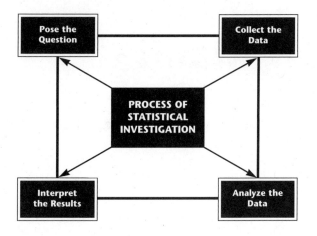

The main purpose of the PCAI model is to give structure and direction to the type of reasoning used in statistical problem solving. (See Investigation 1, *About Us,* for a more detailed discussion of this model.) Reasoning in statistical problem solving involves asking questions such as these:

- Is the question being investigated clearly defined? Is it worthwhile?

- How have the data been collected? Is the sample representative?

- Are the statistics being used the appropriate ones for analyzing the data?

- Do displays of the data present unbiased pictures?

- What conclusions have been drawn in response to the question asked? Do these conclusions make sense?

The exploration of statistics as a content area and the use of the process of statistical investigation work as catalysts for helping teachers and students think about new ways of learning. The process of statistical investigation is centered in inquiry. Student explorations, more often than not, do not have "right" answers. What is valued instead are opinions supported by evidence. Critical thinking and discussion are essential, so teachers focus on developing questioning strategies that will engage students in mathematical discourse.

In their use of the process of statistical investigation, teachers quickly find many ways in which connections across content areas can be made. Because of its rich potential for integration, statistics content is not perceived as an "add on" to an already crowded curriculum. Most importantly, statistics is a topic that is engaging and accessible to all students.

The process of statistical investigation may be visualized through the use of a concept map. The concept map shown on the next page, which has evolved from the Teach-Stat project, incorporates the four components of the statistical-investigation process model (at its center) and elements of statistics content that relate to these components. Related statistical concepts are attached to each part of the process.

The statistical concepts help define what teachers and students need to *know* about statistics. The process of statistical investigation defines what they need to be able to *do.*

We *pose the question* because we want to solve a problem. The problem may involve describing a set of data, summarizing what we know about a set of data, comparing or contrasting two or more data sets, or generalizing from a set of data in order to make predictions about the next case or the population as a whole. The question we ask is related to the problem we are studying and requires that certain variables—categorical or numerical, depending on the nature of the question—be counted or measured.

We *collect the data* by identifying the population under study and the methods for collecting the data. When sampling is involved, different types of samples may be considered, including random samples, convenience samples, or a census. Factors of bias, representativeness, and randomness become important at this stage.

Concept Map of the Process of Statistical Investigation

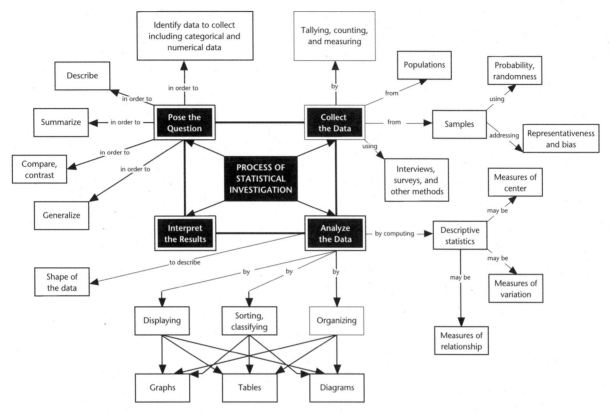

We then *analyze the data,* which may include describing and summarizing the variation in a single set of data, comparing or contrasting two or more sets of data, or examining the association of data collected about two variables. Methods of analyzing data include organizing, sorting, classifying, and displaying data using tables, diagrams, and graphs; and determining descriptive statistics, such as measures of central tendency (mean, median, and mode), measures of variation (range and standard deviation), and measures of association (line of best fit and correlation coefficients).

We then *interpret the results,* which returns us to our original question and the purpose of the investigation. We ask ourselves the question, What do the data we have collected and analyzed say about the question we originally posed?

The Teach-Stat Program

The Teach-Stat program is designed to help elementary-grades teachers learn more about statistics. It also helps them integrate teaching about and teaching with statistics into their instruction. Teach-Stat teachers develop knowledge in several areas, including

- statistics and the process of statistical investigation

- methods for teaching statistics in the elementary grades

- the impact of statistics in our daily lives

- ways to enhance mathematics teaching using a statistical-investigation model

- ways to integrate statistics into science, social studies, and other subject areas

The program consists of four related publications:

Teach-Stat for Teachers: Professional Development Manual is a how-to guide for planning and implementing a three-week teacher-education workshop. It is written in a way that addresses the needs of teachers as learners and teachers' needs for in-service education. Its intended audience is primarily designers of professional-development programs for elementary-grades teachers.

Teach-Stat for Statistics Educators: Staff Developers Manual is a how-to guide for planning and implementing a one-week Statistics Educators Institute. The institute is designed for teachers who, having participated in a three-week program in statistics education and having previously worked with students, will serve as staff-development resource people (statistics educators).

Teach-Stat for Students: Investigations for Grades K–3 and ***Teach-Stat for Students: Investigations for Grades 3–6*** are how-to guides for planning and implementing activities for elementary-grades students that promote the learning of statistics using the process of statistical investigation. The activities provide a variety of "investigation starters," good models that will help teachers to develop their own investigations.

Using *Teach-Stat for Teachers: Professional Development Manual*

The activities in the *Teach-Stat for Teachers: Professional Development Manual* are framed with the Teach-Stat concept map (see page xv) in mind. The four components of the statistical-investigation process define the development of each activity. The activities emphasize components of the process and related statistical concepts. The program has been sequenced to interweave topics in ways that made sense to the developers. See the sample syllabus on pages xxi–xxv for ordering and timing.

The Teach-Stat activities are organized into the following sections:

- **Overview** summarizes the activity

- **Assumptions** details prior knowledge or experience expected in preparation for the activity

- **Goals** specifies goals for activity

- **References** lists related or supporting materials

- **Developing the Activity** organizes the activity according to the four components of the statistical-investigation process: *Pose the Question, Collect the Data, Analyze the Data,* and *Interpret the Data*

- **Summary** reflects on goals of the activity

Appropriate handouts and transparencies follow each activity.

In addition to these statistical-investigation activities, the investigations in this book contain six minilectures and a few activities that offer a general overview of a specific topic. Each minilecture summarizes a major conceptual strand, tying together knowledge developed through the activities and research findings.

We consider the framework in *Professional Standards for Teaching Mathematics* (NCTM 1991) to be central to the teaching and learning of statistics. Implementation of the Teach-Stat program assumes that the professional development provided for teachers models the Standards. As part of the program, participants are asked to read an assignment from the *Professional Standards*.

The Teach-Stat program makes frequent reference to the *Used Numbers* materials *(Measuring: From Paces to Feet; Counting: Ourselves and Our Families; Sorting: Groups and Graphs; Statistics: The Shape of the Data; Statistics: Prediction and Sampling;* and *Statistics: Middles, Means, and In-Betweens)*. We recommend that a set of these materials be made available to teachers for use in planning their work with students (see "References" on page xx).

The Development of the Teach-Stat Program

The original Teach-Stat project involved nine sites of the University of North Carolina Mathematics and Science Education Network. At each of these MSEN centers, a director oversaw the project, and a faculty member acted as site coordinator. Three additional consulting faculty members joined the co-principal investigators and site coordinators and, working with a number of the Teach-Stat teachers, were responsible for planning and implementing the program.

The goal of the project was to develop and implement a program of professional development for teachers of grades 1 through 6 and a staff-development program to prepare statistics educators to serve as resource people equipped to offer the original Teach-Stat professional-development program to other elementary-grades teachers. The project culminated in the publishing of the four Teach-Stat publications, permitting others to implement similar professional-development programs.

During the first year of the Teach-Stat project, each of nine faculty selected six or seven teachers from their regional sites to serve as a pilot team; 57 teachers participated in the first workshop. The three-week workshop (which resulted in the *Teach-Stat for Teachers: Professional Development Manual*) was offered as a residential program at a central site. The faculty worked in teams of three, providing various parts of the program and working with their teams throughout the workshop. Each faculty member met with and visited his or her teacher team throughout the following school year, exploring with the teachers what it meant for them to teach statistics and to integrate statistics content with other subject areas.

In the second year, each of the nine sites offered a revised, nonresidential version of the three-week professional-development program to 24 teachers. The faculty member and the pilot team of six or seven teachers planned and delivered the workshop. By the second summer, faculty and teachers had developed such a good working relationship that the model of a "professional-development team" emerged naturally. The Year 2 participants were able to learn from the Year 1 teachers, who had spent the previous year implementing the program and had many practical classroom examples to share. The Year 1 teachers received a great deal of support, informal how-to-be-a-staff-developer training, and coaching and mentoring from their faculty leaders.

In Year 3, 84 teachers who participated in Years 1 or 2 were selected to become statistics educators. They participated regionally in the equivalent of a one-week institute that focused on the how-tos and whys of staff development (resulting in the *Teach-Stat for Statistics Educators: Staff Developers Manual*). The statistics educators were responsible for developing and delivering a two-week Teach-Stat summer workshop for a set of 24 new teachers at each of nine sites. The final teams of statistics educators varied in composition: some included only Year 1 teachers, some included a balance of Year 1 and 2 teachers, and some included a majority of Year 2 teachers.

From this project emerged a model for helping teachers teach teachers:

- Participate in the workshop
- Implement the program with students
- Assist in the teaching of a workshop
- Offer professional-development programs for other teachers

There are variations to this model. For example, Year 1 teachers benefitted from an informal set of staff-development experiences that occurred while working with the faculty member as they prepared to teach Year 2 teachers. Those Year 1 teachers who became statistics educators then participated in the more formalized set of learning experiences that focused on "teaching teachers" in Year 3, before having a second opportunity to be part of teaching in a Teach-Stat workshop. In cases in which Years 1 and 2 teachers were mixed on a regional team, the teaming of a Year 1 with a Year 2 teacher created a mentor/coach arrangement that supported the newer teacher in his or her initial experiences with teaching other teachers.

We have been delighted and pleased by the reception the Teach-Stat program has received from teachers. We have seen that teachers *do* translate their own learning experiences to learning experiences for their students. Students in many classrooms are actively engaged in exploring statistical ideas throughout the year, not only in their mathematics programs but social studies, science, and other content areas. Teach-Stat teachers report that the process of statistical investigation serves as an excellent tool for organizing interdisciplinary instruction.

One Teach-Stat Success Story

Wendy Rich teaches grade 5 at Seagrove Elementary School in North Carolina. She completed her undergraduate degree, specializing in intermediate education (grades 4 through 6), at the University of North Carolina and joined the Teach-Stat project in its first year (having just completed her second year of teaching).

Wendy didn't begin with an interest in teaching mathematics. "In my first year of teaching, I hated teaching math with a passion. It was a chore; every day—get out your math books. We'd try to struggle through. . . .When the Teach-Stat project came up, I was a new person. I have been improving ever since!"

The Teach-Stat program has had an impact on Wendy's teaching that goes beyond learning new content and ways to address teaching statistics. When she thinks about what she has gained from the project, she notes, "One of the first things would be confidence in my teaching and in [teaching] mathematics."

In talking with Wendy and other teachers in the project, we can hear their awareness of changes in the ways they work with students. "I really have become good at questioning, not leading the students too much, letting them take more of the [leadership] role," Wendy explains. "I find I am probing more to see if they're understanding or to see if they're just giving me an answer. The thinking [that Teach-Stat] has helped me to promote with my students is one of the best ways the project has helped me."

Among the many activities Wendy implemented in her classes was an investigation to introduce her students to the use of histograms. Wendy challenged her students to see how many times they could jump rope without stopping. "We spent time making sure that we defined the problem. What did jumping rope mean? With our feet together or apart? Who turns the rope—the student or other students? Do we get any trial jumps?" She involved both of her grade 5 mathematics classes in the jump-rope investigation and had them compare their results.

The jump-rope data were quite spread out, so there was a need to group the data to represent them. Students first chose to use stem-and-leaf plots, grouping the data in intervals of 10. (See the stem plot on the next page.)

Wendy then introduced her students to histograms, tying them to the structure of the stem plot. "They took off! I was about ready to say they should make their histograms grouped by 10s. For some reason, I hesitated. Then one student volunteered that she was going to group by 25s. With that, different groups of students tried using different intervals—10s, 20s, 25s, 50s. One group used intervals of 100; their classmates thought this was quite clever—they only had to make three bars to show the distribution of the data!" (See the histograms on the next page.)

Number of Times Jumping Rope		
Class 1		Class 2
8 7 7 7 5 1 1	0	1 1 2 3 4 5 8 8
6 1 1	1	0 7
9 7 6 3 0 0	2	3 7 8
5 3	3	0 3 5
5 0	4	2 7 8
	5	0 2 3
2	6	0 8
	7	
9 8 0	8	
6 3 1	9	
	10	2 4
3	11	
	12	
	13	
	14	
	15	1
	16	0 0
	17	
	18	
	19	
	20	
	21	
	22	
	23	
	24	
	25	
	26	
	27	
	28	
	29	
	30	0

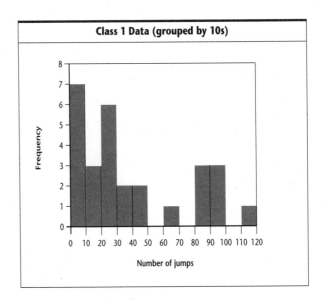

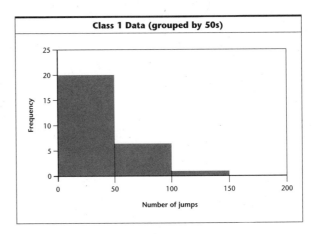

Wendy's students reached several conclusions from their data. "We spent time comparing the two classes' data. We concluded that the reason one class had a higher mean was that there were some outliers—for example, 300, 160, and 151 jumps—that affected it. That got us looking at the median of the data as a way to compare the two classes."

Wendy has seen numerous benefits for her students from this project. "They know the material extremely well, and they have developed real confidence about their abilities to do mathematics. Last year, I started with data collection at the beginning of the year. Not only was it a good bonding experience; it set the stage for the rest of the year. Students were confident; they excelled in making graphs and in analyzing and interpreting data they collected. This carried on throughout the year, and not just in mathematics but in all our work. They *believed* they could be successful!"

References

Corwin, R., and S. Friel. *Statistics: Prediction and Sampling*. Palo Alto, California: Dale Seymour Publications, 1990.

Corwin, R., and S. J. Russell. *Measuring: From Paces to Feet*. Palo Alto, California: Dale Seymour Publications, 1990.

Friel, S., J. Mokros, and S. J. Russell. *Statistics: Middles, Means, and In-Betweens*. Palo Alto, California: Dale Seymour Publications, 1992.

Graham, A. *Investigating Statistics: A Beginner's Guide*. London: Hodder and Stoughton, 1990.

Graham, A. *Statistical Investigations in the Secondary School*. Cambridge: Cambridge University Press, 1987.

Moore, D. *Statistics: Concepts and Controversies*. 3rd ed. New York: W.H. Freeman, 1991.

National Council of Teachers of Mathematics. *Professional Standards for Teaching Mathematics*. Reston, Virginia: National Council of Teachers of Mathematics, 1991.

Russell, S. J., and A. Stone. *Counting: Ourselves and Our Families*. Palo Alto, California: Dale Seymour Publications, 1990.

Russell, S. J., and R. Corwin. *Sorting: Groups and Graphs*. Palo Alto, California: Dale Seymour Publications, 1990.

Russell, S. J., and R. Corwin. *Statistics: The Shape of the Data*. Palo Alto, California: Dale Seymour Publications, 1989.

SYLLABUS
TEACH-STAT Workshop
Three-Week Syllabus

This three-week syllabus is adapted from the syllabus used in the Year 2 summer workshops. Related topics—such as discussions on equity and teacher leadership—are optional; they are included to address issues the project staff felt were important.

The Teach-Stat program may be modified and adjusted for various time frames, such as a two-week workshop, a one-week workshop, or a two-to three-hour awareness workshop. Shortening the program, of course, means eliminating statistics content. Careful attention to identification and clarification of the goals of any professional-development experience is necessary when designing a program of any length.

In this syllabus, it is assumed that participants have been provided with the resources mentioned and that certain videotapes have been purchased for the workshop; see "Resources" at the end of the syllabus.

Week 1

Day 1

8:45–9:15	Check-in and registration
9:15–10:30	Overview of workshop and review of syllabus Introduction and other warm-ups View *Used Numbers* videotape (middle level)
10:30–10:45	Break
10:45–12:00	Concept: The process of statistical investigation **Investigation 1:** *About Us*
12:00–1:00	Lunch
1:00–2:15	Each team presents results and dilemmas Process of statistical investigation is synthesized from the discussions
2:15–3:00	Registration for graduate credit and CEUs
Assignment	Review the *Professional Standards for Teaching Mathematics* (NCTM 1991), pages 1–67 (participants read this material before Day 2 and have responded to the set of questions in Handout 5.1) Discuss *About Us* miniprojects

Day 2

9:00–10:30	Concept (analyze data): Sorting and classifying **Investigation 2:** *Sorting People: Who Fits My Rule?* **Investigation 3:** *Yekttis* **Investigation 4:** *Sorting Things!*
10:30–10:45	Stretch break
10:45–12:00	**Investigation 5:** *Restructuring Mathematics* Participants meet in teams and clarify their understanding of readings through a discussion of Vignette 2.3 from *Professional Standards for Teaching Mathematics* (NCTM 1991)
12:00–1:00	Lunch
1:00–1:45	Debriefing: Sorting and classifying with students—What are the implications for teaching?
1:45–3:00	Concept (analyze and interpret data): Using graphs to picture and interpret data (line plots, bar charts) **Investigation 6:** *Shape of the Data: Using Line Plots* **Investigation 7:** *Shape of the Data: Line Plots to Bar Graphs*
Assignment	Bring in one or more graphs from a newspaper Distribute *Sorting: Groups and Graphs* to participants to review

Day 3

9:00–9:30	Sharing of graphs from newspapers, tying them to a discussion of the shape of data
9:30–12:00	Concept (collect data): Using measurement **Investigation 8:** *Giant Steps, Baby Steps* **Investigation 9:** *What Is the Typical Foot Length of Our Group?* **Investigation 10—Minilecture:** *Children and Measurement*
12:00–1:00	Lunch
1:00–3:00	**Investigation 11:** *Accuracy in Measurement* Debriefing: What are the implications for teaching? View *Used Numbers* videotape (primary level) (optional) **Investigation 12:** *How Close Can You Get to a Pigeon?*
Assignment	Distribute *Measuring: From Paces to Feet* and *Counting: Ourselves and Our Families* to participants to review

Day 4

9:00–12:00	Concept (pose question): Posing the question **Investigation 13:** *Family Size* Discussion of Teach-Stat concept map
12:00–1:00	Lunch
1:00–3:00	Concept (analyze data): Measures of center—median and mode **Investigation 14:** *Median: More Than Just the Middle of the Data* **Investigation 15—Minilecture:** *Types of Data* **Investigation 16:** *About Us Revisited*

Assignment	Distribute *Statistics: The Shape of the Data* to participants to review

Day 5

9:00–12:00	Concept (analyze and interpret data): Using graphs to picture and interpret data (stem plot, histogram) **Investigation 17:** *Shape of the Data: Using Stem Plots* **Investigation 18:** *Shape of the Data: Stem Plots to Histograms*
12:00–1:00	Lunch
1:00–3:00	**Investigation 19:** *Curriculum Overview*
Assignment	**Investigation 20:** *Shape of the Data: Problem Solving*

Week 2

Day 1

9:00–12:00	Concept (analyze and interpret data): Using graphs to picture and interpret data **Investigation 21—Minilecture:** *Graphing* **Investigation 22:** *Raisins Revisited*
12:00–1:00	Lunch
1:00–3:00	Topic: Are girls shortchanged in mathematics education?

Day 2

9:00–12:00	Concept (analyze and interpret data): Using graphs to picture and interpret data; measures of variation; comparing data sets **Investigation 23:** *How Do We Grow?* **Investigation 24:** *Raisins Revisited Revisited* **Investigation 25:** *Graphing Data Using Computers*
12:00–1:00	Lunch
1:00–3:00	Concept (analyze data): Measures of center (mean) **Investigation 26:** *Family Size Revisited* **Investigation 27:** *Cats*

Day 3

9:00–12:00	Concept (analyze data): measures of center (i.e., mean) **Investigation 28:** *Technology* **Investigation 29:** *Building the "Rule" for Finding the Mean* **Investigation 30:** *Means in the News* Discussion of the Teach-Stat concept map Concept (analyze data): Key ideas related to analyzing the data: measures of center and displaying and organizing data **Investigation 31:** *Comparing Sets of Cereal Data*

12:00–1:00	Lunch
1:00–3:00	Team-planning time to revise *About Us*
Assignment	Distribute *Statistics: Middles, Means, and In-Betweens* to participants to review

Day 4

9:00–12:00 **Investigation 32—Minilecture:** *The Mean*
 Investigation 33—Minilecture: *Classroom Assessment of Mathematics*

12:00–1:00 Lunch

1:00–3:00 Continue with minilecture and discussion of assessment
 Have participants try making up different kinds of tasks to assess concepts learned in Investigations 1–33.

Day 5

9:00–12:00 **Investigation 34—Minilecture:** *Linking Probability and Statistics*
 Concept: Actual and expected values in probability
 Investigation 35: *True-False Test*
 Investigation 36: *Removing Markers from a Number Line*
 Investigation 37: *What Are the Odds?*

12:00–1:00 Lunch

1:00–3:00 **Investigation 38:** *Fair Games*

Week 3

Day 1

9:00–12:00 Concepts: Randomness, sampling, and variation
 Investigation 39: *What's in the Bag?*

12:00–1:00 Lunch

1:00–3:00 **Investigation 40:** *Choosing Samples*

Day 2

9:00–12:00 Discussion of the Teach-Stat concept map—key ideas related to collecting data: sampling, ways of gathering data
 Concept (analyze data): Bivariate data
 Investigation 41: *How Tall Are You?*
 Investigation 42: *Are You a Square?*

12:00–1:00 Lunch

1:00–3:00 Principals' session
 General session: Overview of Teach-Stat program
 Activity: How long have you been an educator (half-time or more)? Collect and display responses to the questions on a line plot. Discuss problems and issues raised. Look at the results together as a group.

View *Used Numbers* videotape to see what goes on in classrooms that involve students learning statistics (middle level).
Teacher panel (testimonials; needs for support explained)

Day 3

9:00–12:00	**Investigation 43:** *From Footprint to Stature* **Investigation 44:** *Cats Revisited*
12:00–1:00	Lunch
1:00–3:00	Topics: What do you expect of yourself in using statistics with your students? Discussion of the Teach-Stat concept map

Day 4

9:00–12:00	Open for catch-up
12:00–1:00	Lunch
1:00–3:00	Time to finish Investigation 1: *About Us* projects and prepare to present them
Assignment	Teams present their *About Us* projects

Day 5

9:00–12:00	Reports from teams on their *About Us* projects (see Investigation 1) Award Certificates of Participation

Resources

AIMS Education Foundation. *Jaw Breakers and Heart Thumpers, Grades 3–4*. Fresno, California: AIMS Education Foundation, 1987.

Corwin, R., and S. Friel. *Statistics: Prediction and Sampling*. Palo Alto, California: Dale Seymour Publications, 1990.

Corwin, R., and S. J. Russell. *Measuring: From Paces to Feet*. Palo Alto, California: Dale Seymour Publications, 1990.

Friel, S., J. Mokros, and S. J. Russell. *Statistics: Middles, Means, and In-Betweens*. Palo Alto, California: Dale Seymour Publications, 1992.

Mathematical Sciences Education Board. *Measuring Up*. Washington, D.C.: National Academy Press, 1993.

Moore, D. *Statistics: Concepts and Controversies*. 3rd ed. New York: W. H. Freeman and Co., 1991.

National Council of Teachers of Mathematics. *Professional Standards for Teaching Mathematics*. Reston, Virginia: National Council of Teachers of Mathematics, 1991.

Russell, S. J., and A. Stone. *Counting: Ourselves and Our Families*. Palo Alto, California: Dale Seymour Publications, 1990.

Russell, S. J., and R. Corwin. *Statistics: The Shape of the Data*. Palo Alto, California: Dale Seymour Publications, 1989.

Russell, S. J., and R. Corwin. *Sorting: Groups and Graphs*. Palo Alto, California: Dale Seymour Publications, 1990.

Stenmark, J. K., ed. *Mathematics Assessment: Myths, Models, Good Questions, and Practical Suggestions*. Reston, Virginia: National Council of Teachers of Mathematics, 1991.

Used Numbers (Primary and Middle Level) videocassettes are produced by and may be ordered from Dale Seymour Publications.

INVESTIGATION 1

About Us

Overview

Teams of three to six participants each frame a question they would like to investigate about the entire group of participants. This question becomes the basis for a mini–research project that teams complete during the workshop. Participants use the process of statistical investigation to complete an initial investigation of their questions, reporting preliminary findings to the whole group.

Assumptions

Participants have different backgrounds with respect to their knowledge about and experience with statistics and the process of statistical investigation. While they may be uncomfortable with the open-endedness and apparent lack of structure of the activity, engaging in an activity that requires them to impose their own structure encourages them to develop an understanding of their current knowledge in this area. This process of inquiry promotes discourse and the development of a mathematical community.

Goals

Participants explore the process of statistical investigation. In particular, they

- develop an overview of the process of statistical investigation: posing a question, collecting the data, analyzing the data, and interpreting the results

- recognize the need for refining or redefining the problem situation as part of the process of statistical investigation

- demonstrate current knowledge of key statistical concepts: representations, measures of center, measures of variation, and sampling

Teacher Notes

This is always enjoyed as a beginning learning activity.

Because participants may at first experience a high degree of anxiety, it may be helpful to choose groups carefully. Allow people to be with whom they are most comfortable.

Materials

Pencils, pens, graph paper, lined paper, scissors, colored markers, calculators, colored stick-on dots, chart paper

Transparencies 1.a, 1.b

Handout 1.1

Sample Questions

What kind of coffee do you drink?

What are your hobbies?

How much television do you watch?

How much do you read? What do you read?

How do you get to work?

If you weren't an educator, what field would you be in?

Where were you born?

Teacher Notes

Teachers may want to informally check among themselves that each team is investigating a different question, although investigating the same question can be profitable, too!

Reference

Rosebery, A. S., B. Warren, R. Corwin, A. Rubin, F. Harik, F. R. Conant, and S. Friel. *Reasoning About Data: A Course in Statistical Inquiry for Teachers of Middle School, Grades 5–9.* Cambridge, Massachusetts: Bolt Beranek and Newman, 1990.

Developing the Activity

The development of the activity encompasses the four interrelated components of the process of statistical investigation.

Pose the Question

In small teams of three to six people, participants consider the following problem situation as it pertains to all members of the workshop:

What would you like to know about yourselves as a group?

Each team selects a question they would like to investigate about the group (for example, education level and major, amount of reading done each week, or distance traveled to the workshop).

Collect the Data

After identifying their questions, each team develops or invents their own methods for collecting the data. Depending on the size of the group, a variety of issues may surface.

- Did you ask the same person I did?

- If the group is much larger than 30 people, do we have to ask all of the people our question?

- Our question raises lots of questions about definitions—what do we mean by "How much reading do you do?"

Encourage discussion, but let each team make decisions about their actions without access to the advice of experts in the field.

Analyze the Data

To report their findings, each team should represent their data graphically. They may make use of such strategies as numerical representations, measures of center, measures of spread, symmetry, and observations about the shape of the distribution (clusters, gaps, unusual values) and patterns that emerge. Discussion among team members as they analyze their data is invaluable in highlighting the process of statistical investigation and the components involved.

Interpret the Results

Teams may draw conclusions and raise a number of questions in the process. It is likely that each team will have additional ideas for carrying out their investigations in ways other than those first suggested by their initial questions. This is an important part of the process of statistical investigation and, in this instance, forms the foundation for continuing this activity as a mini–research project to which teams return later in the workshop to complete.

Summary

Each team presents their findings to the group of participants. As each team presents its results, encourage the team and the group to articulate issues, questions, and observations about the statistical investigation process. Note their ideas on the board.

Following the presentations, review the issues and observations that were raised and group them according to the corresponding process of statistical investigation. Use Transparency 1.a to introduce the four components, letting participants decide where each question or issue belongs.

- *Pose the problem.* Are problems framed so that it is clear what information is being sought? What is the purpose for collecting the data: to describe, to summarize, to compare, or to generalize by predicting information about the next case or the population?

- *Collect the data.* How were the data collected?

- *Analyze the data.* What different ways were used to report the data (for example, table, line plot, bar graph, summary statistics)?

- *Interpret the data.* Return to the original purpose for data collection: to describe, to summarize, to compare, or to generalize. Discuss representativeness of data, including sample size. What other questions come to mind based on these data?

Use Transparency 1.a and 1.b to clarify concepts related to each component of the process. Distribute Handout 1.1.

Teacher Notes

It's a good idea to post these graphs and representations to return to throughout the workshop.

Teacher Notes

Samples of the kinds of issues that might surface

People didn't know what was meant by family *so we got different data from people.*

Was any person sampled twice?

Was everyone asked?

Are the data biased?

Are there other ways to report the data?

Is our group representative of a larger population? How can we tell?

Is there a relationship between type of coffee consumed and gender?

About Us Project

As a follow-up, teams reconsider their question(s)—they may even change the question(s)—and complete the process again, preparing a final presentation for the last day of the workshop. Note that in the three-week syllabus, there is time allocated for this as well as for meeting in teams the day before presentations to complete planning.

The Process of Statistical Investigation

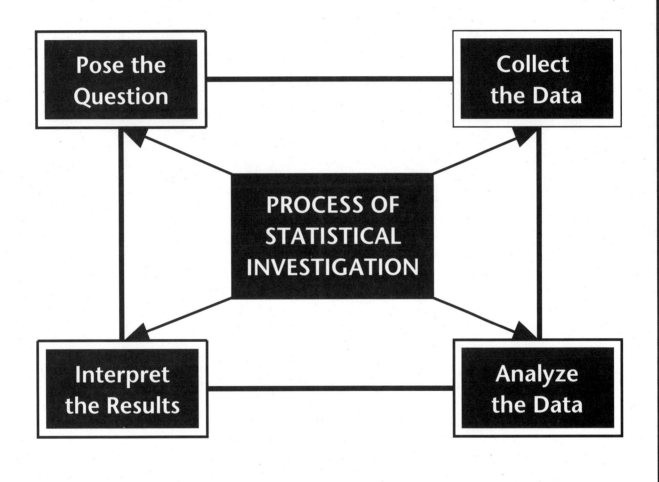

TRANSPARENCY 1.b

Concept Map of the Process of Statistical Investigation

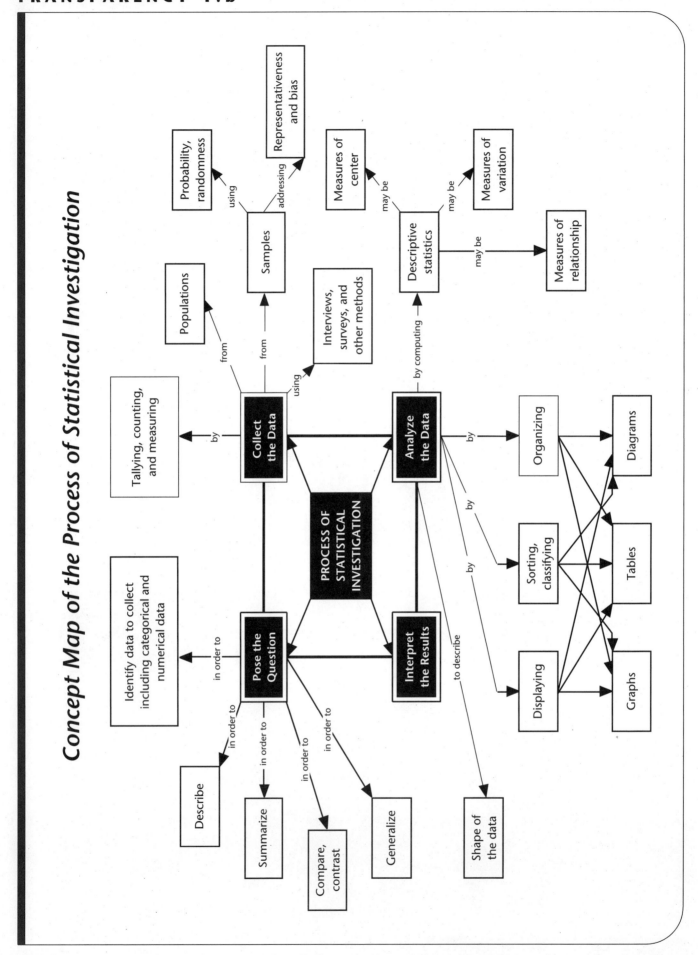

The Statistical Investigation Process

Introduction

The society we live in today is highly technological. Our increased ability to collect, store, and easily retrieve data is changing our world into one in which people are exposed to statistics daily. Numbers, charts, and graphs in newspapers, magazines, advertisements, and on the radio and television are commonplace. Salespeople show us charts and graphs to persuade us to buy their products. Insurance representatives quote such numbers as averages and life expectancies to convince us of the need for insurance. Medical breakthroughs are the result of carefully researched studies; the popular media often report the latest studies in medical journals. We are sometimes confronted with reports of false uses of data.

To be successful in this society, we must develop our "data sense." Data sense includes being comfortable with our abilities to pose questions, collect and analyze data, and interpret the results in an effort to answer the original questions. However, it extends well beyond this to include comfort and competence in reading and listening to the reports in the mass media that are based in statistics.

Reading and listening with data sense is more than understanding graphs and statistics. It involves evaluating the statistical investigation process used to explore these questions, including assessing

- the questions asked; for example, Are the questions clearly defined? Are they worthwhile?

- the data-collection strategies; for example, How were the data collected? Is the sample representative?

- the analysis and presentation of the data; for example, Are the statistics used the statistics that should be used? Do the displays of the data present an unbiased picture of the data?

- the interpretation of the data; for example, What conclusions are drawn? Do these conclusions make sense?

What Is Statistics?

Statistics is the science that relies on data to answer questions. What makes a problem a statistical investigation is the way the question is asked, the role and nature of the data, the ways the data are examined, and the types of interpretations made from the examination. A statistical problem typically contains four components, in some order: (1) question, (2) data, (3) analysis, and (4) interpretation. This four-point framework is a foundation for all Teach-Stat activities.

General Strategies for Statistical Investigation

A statistical investigation involves four stages in a model—the PCAI model (Graham, 1987):

- *Posing the question* Identifying a specific question to explore and deciding what data to collect to address the question

- *Collecting the data* Deciding how to collect the data as well as collecting the data

- *Analyzing the data* Organizing, summarizing, describing, and displaying the data, and looking for patterns in the variation of the data

- *Interpreting the results* Using the results from the analyses to make decisions about the original question

The PCAI model gives structure and direction to the reasoning used in statistical problem solving. The individual components in the model are not necessarily self-contained, and the process does not always follow a particular sequence.

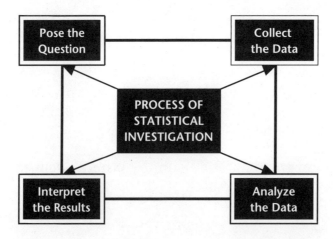

Posing the Question

Every statistical investigation begins with a problem. Ideally, the problem has some relevance to students' lives. As much as possible, students should be allowed to investigate their own problems.

Relevance is also influenced by curricular and instructional goals, which may motivate questions that are not posed by the students. Activities should encourage intuitive problem solving through exploration. Interdisciplinary connections are viable as part of the effort to identify applicable, relevant problems.

The initial statement of the problem may be vague, and consequently the first step in the process is to *pose a more focused and precise statement of the problem*—that is, to refine the question. Question refinement may occur more than once, until a question is formulated that may be addressed using data.

The measurements required to address the question should be identified at this stage. Both the final form of the question and the types of variables have an effect on how the data are collected, the type of analysis performed, and the interpretation of the results.

It is essential that students (and teachers) have a clear understanding of the relationship between the problem under study, the specific questions that will be asked, and the variables considered in addressing these questions. Question-formulation skills are best developed through experience.

Collecting the Data

The heart of statistics is data. Data are not just a set of numbers; they are a set of related measurements. The types of conclusions that can be made from the data depend on the data-collection design. How to organize the data collection depends on the nature of the question(s) and variables that have been identified.

For many investigations, data is collected from all members of a group. Other questions may require that a sample from the group be selected. In this case, notions of fairness, randomness, and sample size need to be addressed. Cause-and-effect questions must be addressed by collecting data from controlled or comparative experiments. Recording data and measurement errors is also important. Attention must be given to identifying the many sources of variation in the measurements (for example, inherent differences in the units being measured, the measurement process, and randomness).

HANDOUT 1.1

As often as possible, students should collect their own data. This is especially true for elementary-grades students. The use of existing data may be appropriate for comparative purposes; however, manufactured data should be avoided.

To explore the typical time people go to sleep at night, a number of teacher educators at a recent conference were asked, What time did you go to sleep last night?

Times Teacher Educators Went to Sleep					
11:00	2:00	12:30	12:50	11:30	11:40
10:25	12:15	10:30	11:30	11:20	8:00
10:15	12:30	10:30	10:30	10:30	10:15
12:00	11:30	11:30	9:30	11:00	1:30
11:00	10:45	12:15	12:00	10:45	11:00
12:05	11:30	12:15	12:00	12:00	11:30
1:00	11:30	11:30	9:30	11:00	10:30

The teacher educators refined the question by asking additional questions: How did you decide what time you went to sleep? Did you have any problems answering this question?

What kinds of issues would be raised if *you* tried to clarify the question?

Analyzing the Data

Data analysis focuses on ways to represent data for the purpose of identifying patterns of variation in the data. Students must be offered opportunities for exploration and should be encouraged to create their own representations. Often such representations provide insight into their understanding of the question at hand and how the data relate to that question. The appropriate representations of the data depend on the nature of the question(s) and the type of data collected. That is, *how* the data are represented depends on *why* the data have been collected and *what* type of data has been collected.

Following are some ways these data might be represented.

HANDOUT 1.1

What time did we go to sleep last night?

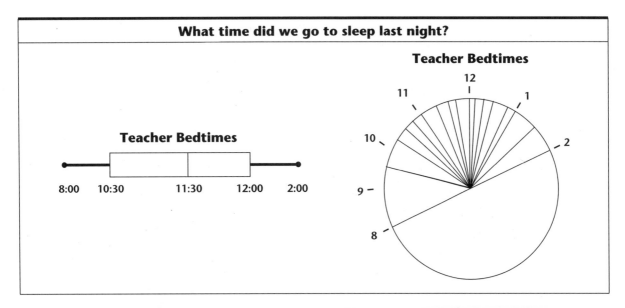

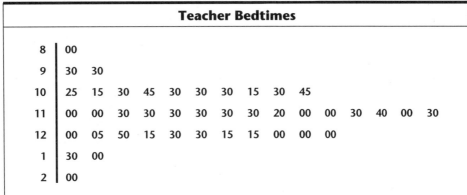

Teacher Bedtimes

8	00													
9	30	30												
10	25	15	30	45	30	30	30	15	30	45				
11	00	00	30	30	30	30	30	30	20	00	00	30	40	00	30
12	00	05	50	15	30	30	15	15	00	00	00			
1	30	00												
2	00													

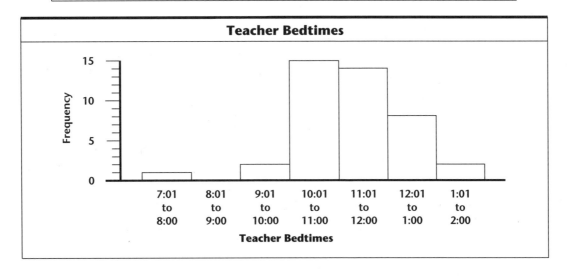

Statistical representations of data include graphical displays and numerical summaries. Graphical displays provide visual descriptions of variability in data. Do the data cluster in places? Are there gaps or unusual values (outliers)? Statistical concepts such as typical value (most common, center); spread, dispersion, and variability; and shape are easily seen by looking at one or more types of graphical displays. Numerical summaries provide a means for summarizing the conceptual ideas conveyed through graphical displays.

Interpreting the Results

How do the results from the data analysis relate to the original problem? Interpreting the results requires making sense from the analysis to address the questions being asked. Because the interpretation must be communicated verbally or in writing, results from the analysis must be translated from graphical displays and numerical summaries into meaningful statements in English.

The following questions might be asked about how to interpret the results:

- How would we describe the people in this group based on the data we collected?

- If someone walked into this room right now, what time might we predict that person went to sleep last night? Why?

- Based on our data, can we generalize to all people attending the conference? Why or why not?

- What can we say about the ways people chose to represent the data? How well do the displays help summarize the data? Are there any questions about the representations we used?

- How does what we did relate to the original context? What might we want to ask now to refine our understanding of adults' sleeping behavior?

Students should be encouraged to be skeptical of their results and to ask critical questions that work to pinpoint possible shortcomings in any of the previous stages. Do the questions asked adequately address the problem under study? Are the measurements appropriate for the questions asked? Have the data been collected properly and the measurements recorded accurately? Are the data representations appropriate for the questions asked and the data collected? The strength of any conclusions drawn from the data depends on the answers to these questions. During this stage, rethinking the original questions and identifying new problems or questions may provide fuel for the process to cycle into another statistical investigation.

Concept Mapping the Process of Statistical Investigation

The process of statistical investigation may be depicted with a concept map (see the next page). The center of the map displays the four components of the process, and attached to each process are related statistical concepts.

- We *pose the question* because we want to describe a set of data, summarize what we know about a set of data, compare or contrast two or more sets of data, or generalize from a set of data, making predictions about the next case or the population as a whole.

  The kind of data we collect depends on the nature of the question posed; we need to distinguish among different data types, including categorical and numerical data.

- When we *collect the data*, we identify methods for collecting the data and determining the sample or population that will be considered. When sampling is involved, different kinds of samples may be used, including convenience samples or self-selected samples. During this step, instances of bias, representativeness, and randomness become important.

- When we *analyze the data*, there are three areas of emphasis: (1) describing the shape of the data; (2) computing descriptive statistics including mean, median, mode, range, and correlation; and (3) organizing, sorting, classifying, and displaying the data using tables, diagrams, and graphs.

- When we *interpret the results,* we are lead back to our question and to reflecting on our purpose(s) for the investigation.

References

Crockcroft, W.H. *Mathematical Counts.* London: H.M.S.O., 1982.

Goodchild, S. "School Pupils' Understanding of Average." *Teaching Statistics* 10 (3): 77–81 (1988).

Graham, A. *Investigating Statistics: A Beginner's Guide.* London: Hodder and Stoughton, 1990.

Graham, A. *Statistical Investigations in the Secondary School.* Cambridge: Cambridge University Press, 1987.

Moore, D. *Statistics: Concepts and Controversies.* 3rd ed. New York: W.H. Freeman, 1991.

HANDOUT 1.1

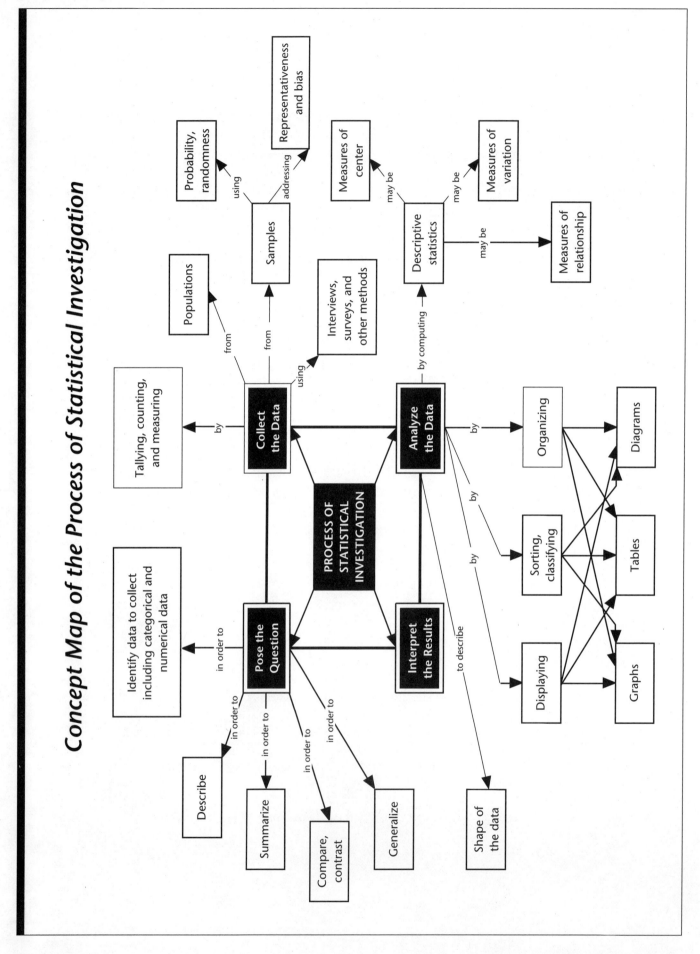

INVESTIGATION 2
Sorting People: Who Fits My Rule?

Overview

In this activity participants observe, classify, count, and record data about themselves.

Assumptions

Participants develop skills in classifying data by identifying a set of attributes and then sorting members of the group based on characteristics of those attributes.

Goals

Participants explore the concept of classification. In particular, they

- make observations about similarities and differences in data, using themselves to compose a group of objects

- demonstrate thinking flexibly about attributes of the data

- use negative information (for example, something is not the case) to clarify the definition of attributes

- use graphical representations to view data that has been sorted and classified

Reference

Russell, S. J., and R. Corwin. *Sorting: Groups and Graphs.* Palo Alto, California: Dale Seymour Publications, 1990.

Teacher Notes

This activity is excellent for introducing the classification of animals in the science curriculum.

Materials

Interlocking cubes, calculators

Teacher Notes

Sample rules

is wearing a watch

has blue eyes

is wearing shoes that tie

is wearing red

is wearing a shirt with buttons

is sitting on the left side of the room

is wearing a belt

Developing the Activity: Part 1

Participants will play the classification game Guess My Rule.

Pose the Question

Look around the room at the people here today. What are some ways we might try to group ourselves? What are some characteristics we can use to form these groups?

Participants may note several ways this can be done using a variety of attributes.

As you may know, when we divide information into groups, we are using an important tool: classification. Classification is used to collect, organize, and interpret data.

Today, I want you to play a game with me. It is called Guess My Rule. I have a mystery rule in mind that you will try to guess.

My rule tells something about some of the people in this room. For example, Sally and Marcelo fit my rule. Let's have them stand over here. Gail and Robert don't fit my rule. We'll ask them to stand over there. Can you guess my rule?

Collect and Analyze the Data

Participants try to guess the rule you are using (for example, "is wearing a watch") by asking such yes/no questions as, "Does [Theresa] fit your rule?" After your response (yes or no), the person mentioned joins one of two groups so participants can see who does and who does not fit the rule. "Wrong" guesses are just as important as "right" guesses. When participants think they can guess your rule, you might have them test their theories by giving another example (that is, choose a person in the room who fits or doesn't fit the rule to *demonstrate* their understanding without giving the rule away). This gives others more time to look for the attribute you have selected, as well as providing an additional piece of data.

Interpret the Results

When enough evidence has been gathered and most participants appear to know the rule, allow them to share the rules they have been using. Encourage them to give reasons for their rules, describing the thinking processes that led them to the rules they selected. Do all the participants state the rule in the same way? Is there more than one rule working, given the data they considered? (For example, you may have been using the rule "is wearing a watch" but your groupings may be such that the rule "is wearing a belt" also fits. These are satisfying "aha!" experiences.)

Play another round of the game, using a new rule. Then, have participants work in small groups to develop a rule of their own. Have two or three groups work with the rest of the participants to play Guess My Rule using rules they have generated.

Summary

Reconvene the participants, and discuss their observations about classification used in this context. What do they anticipate might happen with students they teach? In what other ways might they integrate data classification into their teaching?

Developing the Activity: Part 2

Participants will use interlocking cubes to model the number of people who do and do not fit a given rule.

Pose the Question

We have played several rounds of Guess My Rule and counted the number of people who did or did not fit each of our rules. What was one of the rules?

Participants will recall rules (for example, "is wearing blue").

Collect and Analyze the Data

Using one rule, let's gather data about the people in this room. Each person will receive one interlocking cube. With our cubes, we will build two towers: one of cubes that represent people who fit the rule, and one of cubes that represent people who do not fit the rule. Let's make the two towers now.

Interpret the Results

As a group, discuss what it means to have a tower of cubes that shows the people who fit the rule. Each cube represents one person, so the number of cubes in the tower shows how many people fit the

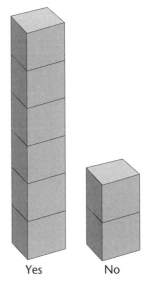

Is wearing blue

rule. Next, consider what it means to have a tower of cubes that shows the people who do not fit the rule. Why might identifying people who do *not* fit the rule be different from identifying people who do fit the rule?

Comparing data is now possible. We can talk about the number of people who fit the rule as being *more than, less than*, or the *same as* the number of people who do not fit the rule. The cube towers provide a way to visualize this information without having all the people stand in one of the two groups. Participants are now moving toward familiarity with more abstract ways of representing data.

Summary

Work with participants as they consider in what ways their students may use surveys as an outgrowth of Guess My Rule activities. Their students can identify questions, gather data, and display it with cubes. They can discuss what their data tell them about the class. Teachers may encourage their students to talk about what attributes are represented with the interlocking cubes.

Teachers will find the development of this activity for student use detailed in *Sorting: Groups and Graphs* (see Reference p. 15).

INVESTIGATION 3

Yekttis

Overview

In this activity participants explore attributes of Yekttis, which are imaginary animals. Playing the game Guess My Rule with cards portraying Yekttis, participants make observations and develop hypotheses about the common characteristics of Yekttis.

Assumptions

Participants have some experience investigating similarities and differences in sets of related objects. They develop skills in classifying groups using particular attributes and in sorting members of the group based on characteristics of the attributes.

Goals

Participants explore the concept of classification. In particular, they

- make observations of similarities and differences using data about a group of objects

- explore ways to change or hold constant attributes of the data

- use negative information to clarify the definition of attributes

- gain experience collecting data and building theories about data

Reference

Russell, S. J., and R. Corwin. *Sorting: Groups and Graphs.* Palo Alto, California: Dale Seymour Publications, 1990.

Materials

Set of large Yektti cards, sets of small Yektti cards, packs of Yektti word cards, one-ring and two-ring Venn diagrams drawn on tagboard, calculators

Handout 3.1

See *Sorting: Groups and Graphs* (Russell and Corwin 1990), pages 25–38, for the Yektti cards.

Two Yekttis

Developing the Activity: Part 1

Participants will learn the story of the Yekttis and, in a whole-group activity, small groups ask questions until they have determined how many Yekttis are in a complete set.

Pose the Question

Read the following story of the Yekttis (from *Sorting: Groups and Graphs*, pages 26–27, used with permission). Let participants know that the story sets the context for the investigation.

Lee and Anita, a pair of eight-year-old twins, discovered some strange creatures near their home in Wyoming. These creatures were living in abandoned prairie dog burrows next to a dirt road that the twins used as a shortcut on their way to school. Lee and Anita started studying these creatures. They visited them every chance they had. Because these creatures never came all the way out of their holes in the ground, Lee and Anita could see only their heads. The creatures looked as though they might have come from another planet.

Lee loved to make up codes and learn about languages. After a few months, he learned how to say some words in the creatures' own language, and he taught them a few words in English and in Spanish. He learned that the creatures called themselves Yekttis (YEK-tees), that they came from a very distant planet, and that they were peaceful creatures.

Anita liked to study different kinds of living things. She decided to do a report about the Yekttis for a science project at school. She made a sketch of the head of each of the Yekttis she had seen. She noticed that a lot of them were similar to each other, but that no two were exactly alike. She used her sketches to figure out how she could describe to other human beings what the Yekttis looked like.

Today you will look at copies of Anita's sketches and see whether you can figure out how to describe the Yekttis.

Display three or four large Yekttis cards on a window ledge or chalk tray so all participants can see. Select cards that show some of the varying characteristics and attributes. For example, you may want to display three cards: at least two of different shapes, at least two with different numbers of antennae, at least two with different eyes, and so on.

Here are a few of the Yekttis that Lee and Anita saw. Remember, they discovered that the Yekttis were similar, but no two were exactly alike. I want you to figure out how many Yekttis there are in all. You can do this by asking me yes/no questions. You will be asking me for a particular Yektti you think I might have. If I have one that satisfies your question, I will show it to you. If not, I will respond that I don't have such a Yektti.

Collect and Analyze the Data, and Interpret the Results

Working in small groups, participants ask yes/no questions as part of a whole-group activity. After several additional Yekttis have been identified, participants meet in their groups to discuss what they now know and what information they still need, if any, to determine how many Yekttis there are in the complete set. Continue by posing further questions. When most participants indicate they know how many Yekttis there are, discuss ways that this information was determined.

Developing the Activity: Part 2

Participants will take turns selecting a rule and answering questions about whether a chosen Yektti does or does not fit the rule.

Collect and Analyze the Data

Divide participants into groups of three. Give each group a set of Yektti cards, a pack of Yektti word cards, and a one-ring Venn diagram. Participants take turns choosing a mystery rule (for example, square) by selecting one of the Yektti words cards, keeping it hidden from the other players.

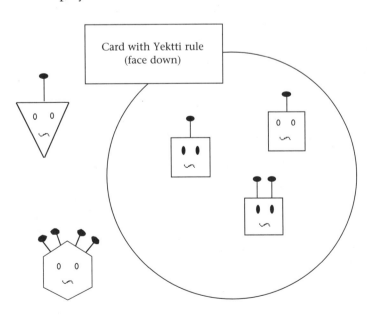

Sample Dialogue

Participants might ask, *Do you have a Yektti with more than four antennae?* Your response would be, *No.*

Do you have a Yektti with a square head [assuming that none is already displayed]*?* Your response would be, *Yes, here's such a Yektti.*

Do you have a Yektti that is a different shape from those that are displayed? Your response would be either *Yes, here's such a Yektti,* or *No,* depending on what is displayed.

Teacher Notes

There are 4 shapes, 2 kinds of eyes, and 4 different numbers of antennae, yielding a total of 4 x 2 x 4 = 32 Yekttis.

Variations

We did this first as a large group using a large plastic hoop taped to the board, then we split into small groups.

Once a rule has been chosen, the rule chooser places the rule card face down on the edge of the one-ring Venn diagram. The others in the group guess the rule by presenting a Yektti card and asking, "Does this Yektti fit your rule?" If the answer is yes, the Yektti card is placed inside the ring. If not, it is placed outside the ring.

After participants gather data, they examine the Yektti cards that fit the rule and those that do not.

Interpret the Results

After examining the Yektti cards, participants draw a conclusion about the rule and check their rule against the rule card.

Developing the Activity: Part 3

Participants will take turns selecting two rules and answering questions about whether a chosen Yektti does or does not fit the rules.

Collect and Analyze the Data

Divide the class into groups of three. Give each group a set of Yektti cards, a pack of Yektti word cards, and a two-ring Venn diagram. Participants take turns choosing two mystery rules, one for each of the Venn diagram rings.

Teacher Notes

The rules in this two-ring Venn diagram are square and ringed eyes.

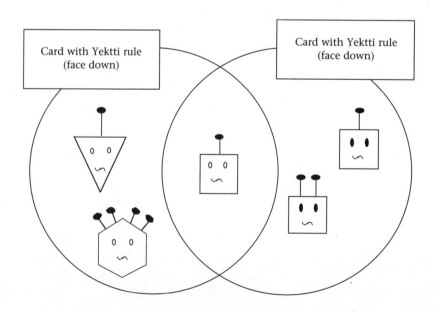

Once rules have been chosen, the rule chooser places the rule cards face down, one on the edge of each of the two rings. The others in the group try to guess the rules by presenting a Yektti card and asking, "Does this Yektti fit your rules?" If the Yektti card fits one of the rules, it is placed inside the correct ring; if it satisfies both rules, it is placed in the intersection of the rings. If it satisfies neither rule, it is placed outside the two-ring Venn diagram.

After participants gather data, they examine and discuss the Yektti cards that fit the rules and those that do not.

Interpret the Results

After examining the cards, participants can draw a conclusion about the rules. They may check their rules against the rule cards.

Summary

Work with participants as they consider in what ways their students may be introduced to sorting and classifying using Yekttis and Venn diagrams. Teachers will find the development of this activity for student use detailed in *Sorting: Groups and Graphs* (see Reference).

Teacher Notes

Be sure to leave enough time for participants to explore this problem situation trying different pairs of rules.

Using Venn Diagrams: Sorting and Classifying Yekttis

One-Ring Game

Divide players into groups of three. Give each group a set of Yektti cards, a pack of Yektti rule cards, and a one-ring Venn diagram. Players take turns selecting a mystery rule (for example, "Yektti is square") by drawing one of the rule cards, keeping it hidden from the other players.

Once a rule has been chosen, the rule chooser places the rule card face down on the edge of the ring. Others in the group guess the rule by presenting a Yektti card and asking, "Does this Yektti fit your rule?" If the answer is yes, the Yektti card is placed inside the ring. If not, it is placed outside the ring.

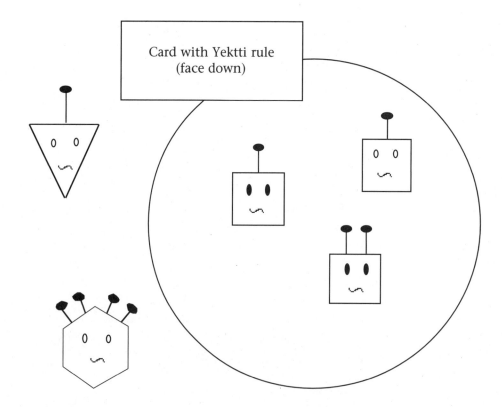

HANDOUT 3.1

Two-Ring Game

Divide players into groups of three. Give each group a set of Yektti cards, a pack of Yektti rule cards, and a two-ring Venn diagram. Players take turns selecting two mystery rules, one for each of the rings in the Venn diagram.

Once the rules have been chosen, the others in the group try to guess the rules by presenting a Yektti card and asking the rule chooser, "Does this Yektti fit your rules?" If the Yektti card fits one of the rules, it is placed inside the correct ring; if it satisfies both rules, it is placed in the intersection of the rings. If it satisfies neither rule, it is placed outside the two-ring Venn diagram.

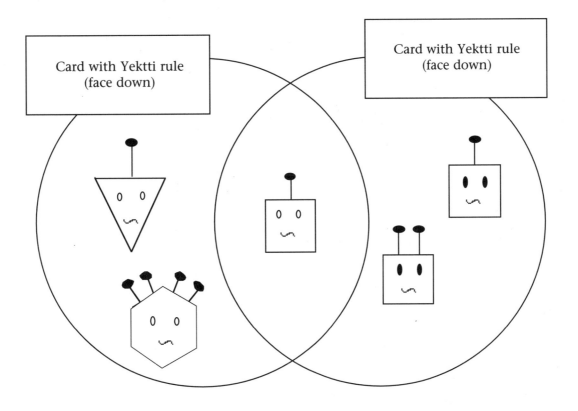

The rules for the above diagram are "ringed eyes" or "square."

INVESTIGATION 4

Sorting Things!

Overview

In this activity participants observe, classify, count, and record data about groups of objects. This activity can easily be integrated with the science curriculum.

Assumptions

This investigation builds on the earlier investigations with the Guess My Rule game and Yekttis. Participants develop skills in classifying groups using particular attributes and in sorting members of the groups based on characteristics of the attributes.

Goals

Participants explore the concept of classification. In particular, they

- construct attributes for use in sorting data

- make observations about similarities and differences across the group of objects in the data

- demonstrate thinking flexibly about attributes of the data

- use negative information (for example, something is not the case) to clarify the definition of attributes

- use graphical representations to display data that has been sorted and classified

- gain experience collecting data and building theories about data

Reference

Russell, S. J., and R. Corwin. *Sorting: Groups and Graphs*. Palo Alto, California: Dale Seymour Publications, 1990.

Materials

Collections of 20 items, one-ring Venn diagrams, two-ring Venn diagrams, calculators

Teacher Notes

Here is an example of a collection of 20 items.

paper clip
button
thumb tack
straw
piece of paper
piece of spaghetti
battery
bar of soap
earring
safety pin
rubber band
spoon
magnet
penny
checker
screw
shoe string
paper cup
scissors
pencil

Sample Dialogue

I'd like you to gather around this table. We're going to play Guess My Rule using this collection of 20 things and the one-ring Venn diagram.

I have a rule that you must guess. It is based on the characteristics of the things we have here. Now I will put two of the things in the one-ring Venn diagram [choose a rule such as "round things"]. You may select other things from the collection, and I will place them either inside or outside of the one-ring Venn diagram. When you think you know my rule, I'll ask you to help me decide where to place items that other participants select.

Now you will return to your tables. Each group has a 20-item collection at their table. Play the one-ring Guess My Rule game using your collection. As a group, discuss several ways that you can classify objects in your collection. When you are ready for a challenge, let me know.

Challenge: *I have two one-ring Venn diagrams. I'm going to use two rules [for example, round, made of wood]. Help me place your 20-item collection in the correct circles using these two rules.*

Developing the Activity

You will need collections of 20 items for this activity. You may want to have several collections of the same set of 20 items so that groups of participants can use the same materials. You may also want collections of different sets of 20 items to motivate different schemes for categorizing and sorting data.

This activity ties together earlier work done with Guess My Rule, a game used to sort people, and the use of Venn diagrams to sort Yekttis. You may want to use both one-ring and two-ring Venn diagrams with participants as tools for labeling and sorting the 20-item collections. Use of one- or two-ring diagrams in classrooms will depend on the developmental level of the students.

Pose the Question

Play the one-ring Guess My Rule game. Identify one attribute of the data (for example, shape) and a value of that attribute (for example, round) as your label for the Venn diagram (see "Sample Dialogue"). When most or all participants can guess the rule, have a few state the rule. Foster discussion that promotes the stating of the rule using different words.

Collect and Analyze the Data, and Interpret the Results

Encourage participants to explore many different ways to group their collections. When you meet with a group, discuss what they have done. Introduce the Two-Ring Game in which they group their 20-things collections into two categories.

Participants will eventually face the dilemma of what to do, for example, with "round things made of wood." This permits introduction of the two-ring Venn diagram; objects that belong in both categories can be placed in the intersection. Once participants understand what is happening, encourage them to play the two-ring game with overlapping rings.

Summary

Participants have been using Venn diagrams to sort and classify data. While their students may not need the following formalization, you may want to discuss ways of naming sorted groups of items.

In the one-ring game, items in the circle are positively labeled (for instance, round), while things outside the circle are negatively labeled (not round), using the identified value (in this example, round, one of many values for the attribute of shape).

In the two-ring game, items in each of the circles are positively labeled (for instance, round, made of wood); items in the intersection are labeled using the connector *and* (round and made of wood); and items outside the circles are negatively labeled using the connector *and* (not round and not made of wood). Finally, the two circles together are positively labeled using the connector *or* (for example, round or made of wood). Work with participants to understand the uses of labels and connectors.

It is important that participants understand the use of the language of the one-ring and two-ring Guess My Rule games. However, it is not necessary that they require similar competence from their students. Teachers' understanding and modeling of correct language will provide background for their students as they play the games.

The concepts explored in this activity provide the foundation for work with sorting and classifying information using databases, particularly computer databases. You may want to return to this work when you explore the *Cats* database so that participants may build the connections from their work here for use in their work with databases.

Teacher Notes

Sorting and classifying are useful skills in other disciplines, including science and social studies. Venn diagrams can be used to sort information in a variety of

INVESTIGATION 5

Restructuring Mathematics

Overview

In this activity participants read and think critically about the *Professional Standards for Teaching Mathematics* (NCTM 1991) and how the instruction called for in the *Professional Standards* differs from the principles of traditional instruction.

Assumptions

Participants have read the *Professional Standards for Teaching Mathematics* (NCTM 1991) pages 1–67 and have completed the pre-workshop written activity related to Vignette 2.3 (see Handout 5.3 and the sample syllabus).

Reference

National Council of Teachers of Mathematics. *Professional Standards for Teaching Mathematics*. Reston, Virginia: NCTM, 1991.

Developing the Activity

Participants will read the first two sections of the *Professional Standards for Teaching Mathematics*:

- *Introduction* (pages 1–18)

- *Standards for Teaching Mathematics* (pages 19–67)

As an overview, you may want to summarize the major sections of the *Professional Standards for Teaching Mathematics*, using the transparencies provided. This will help participants focus their attention on the four major components of teaching:

- Tasks

- Discourse

- Learning environment

- Analysis (reflection)

Teacher Notes

Standards for Teaching Mathematics

Worthwhile Mathematical Tasks

Teacher's Role in Discourse

Student's Role in Discourse

Tools for Enhancing Discourse

Learning Environment

Analysis of Teaching and Learning

Materials

Transparencies 5.1a through 5.1d

Handout 5.1

Teacher Notes

As they monitor students' understanding of, and dispositions toward, mathematics, teachers should ask themselves questions about the nature of the learning environment they have created, of the tasks they have been using, and of the kind of discourse they have been fostering (NCTM 1991).

Have participants work in groups of four or five to discuss their responses to the questions relating to Vignette 2.3. Their discussions should help to clarify their understanding of the four components of teaching.

You may want to spend 10 to 15 minutes at the end of the activity having participants share their reflections about the impact of the *Professional Standards for Teaching Mathematics*.

Summary

This activity is designed to provide an introduction to the *Professional Standards for Teaching Mathematics*. It is intended that *tasks, discourse, learning environment,* and *analysis* or *reflection* become common language when discussing teaching and learning throughout the Teach-Stat workshop.

Guidelines for the Modern Teacher

The most useful metaphor for describing the modern teacher is that of an intellectual coach. At various times, this role will require that the teacher be

- A *role model* who demonstrates not just multiple paths to a solution but also the false starts and higher-order thinking skills that lead to the solutions of problems

- A *consultant* who helps individuals, small groups, or the whole class to decide if their work is "on track" and making reasonable progress

- A *moderator* who poses questions to consider but leaves much of the decision making to the class

- An *interlocutor* who supports students during class presentations, encouraging them to reflect on their activities and to explore mathematics on their own

- A *questioner* who challenges students to do work that is both reasonable and purposeful and who ensures that students can defend their conclusions

Reference

National Research Council. *Counting on You: Actions Supporting Mathematics Teaching Standards.* Washington, DC: National Academy Press, 1991, pp. 13–14.

NCTM Professional Standards for Teaching Mathematics

- Standards for Teaching Mathematics

- Standards for the Evaluation of the Teaching of Mathematics

- Standards for the Professional Development of Teachers of Mathematics

- Standards for the Support and Development of Teachers and Teaching

- Next Steps

Reference

Professional Standards for Teaching Mathematics. Reston, Virginia: National Council of Teachers of Mathematics, 1991.

Assumptions About Teaching Mathematics

- The goal of teaching mathematics is to help all students develop mathematical power.

- *What* students learn is fundamentally connected with *how* they learn it.

- All students can learn to think mathematically.

- Teaching is a complex practice and hence not reducible to recipes or prescriptions.

Reference

Professional Standards for Teaching Mathematics. Reston, Virginia: National Council of Teachers of Mathematics, 1991.

Standards for Teaching Mathematics

- Worthwhile Mathematical Tasks

- Teacher's Role in Discourse

- Student's Role in Discourse

- Tools for Enhancing Discourse

- Learning Environment

- Analysis of Teaching and Learning

Reference

Professional Standards for Teaching Mathematics. Reston, Virginia: National Council of Teachers of Mathematics, 1991.

Using the Professional Standards for Teaching Mathematics

Please read pages 1–67 in the *Professional Standards for Teaching Mathematics* (NCTM 1991). (We know this may sound like a lot of reading, but we think you will enjoy the vignettes.) Study Vignette 2.3 on pages 40–42, and write your responses to the following questions.

1. How did the teacher set up the learning environment? How is this class's climate different from that of traditional classes? [Standard 5]

2. How was the task structured to facilitate the teaching of significant mathematics concepts? [Standard 1]

3. What types of questions did Mr. Luu ask the students? How did he encourage discourse? [Standard 2]

HANDOUT 5.1

4. When during the discussion did Mr. Luu let students struggle with their understanding of the concepts? [Standard 2]

5. How did Mr. Luu use tools such as concrete materials, diagrams, models, analogies, and arguments to enhance discourse? [Standard 4]

6. Examine the goals applying to the discourse of students listed on page 45. Which of these roles were evident in Vignette 2.3? [Standard 3]

7. How did Mr. Luu alter his instruction based on information he gained regarding his students' understanding? [Standard 6]

8. What implications from the *Professional Standards for Teaching Mathematics* apply to your students' grade level?

INVESTIGATION 6
Shape of the Data: Using Line Plots

Overview

Participants explore two different problems in order to practice collecting data and displaying it using line plots. The focus of their analysis is the shape of the data and their interpretation of how they can use what they know about the data to respond to the question, What do you expect will happen to the shape if we collect more data? The two problems are

- How many raisins are in a half-ounce box?
- What are the dates on the pennies you have with you now?

Assumptions

Participants have had little or no introduction to data analysis. They may have participated in Investigation 1, *About Us*.

Goals

Participants explore the concept of the shape of the data. Formally, this includes consideration of five components:

- symmetry or skewness
- presence or absence of single or multiple peaks
- center(s) of the data
- degree of spread around the center
- deviations from the regular pattern in the data, including gaps and outliers

Informally, we begin by talking about clumps, bumps, holes, and what's typical about the data, rather than using the terms above.

The line plot is introduced as a way to display data.

Materials

Colored stick-on dots or stick-on notes, chart paper, blank transparencies, transparency markers, colored markers, individual (half-ounce) boxes of raisins, chart paper or blank transparency (for line plot), calculators

Transparencies 6.a through 6.d

Handout 6.1

References

Russell, S. J., and R. B. Corwin. *Statistics: The Shape of the Data*. Palo Alto, California: Dale Seymour Publications, 1989.

Landwehr, J. M., and A. E. Watkins. *Exploring Data*. Rev. ed. Palo Alto, California: Dale Seymour Publications, 1995.

Developing the Activity: Part 1

Participants will count the raisins in a sample of half-ounce boxes (one box for each participant), record and organize the results, and describe the shape of the data distribution.

Pose the Question

Give a box of raisins to each participant. Ask them to keep the boxes closed.

Does anybody have an idea about how many raisins are in a half-ounce box of raisins?

Let participants make an estimate and then open their boxes so they can see the layer of raisins at the top of the box.

What do you think now? Do you want to revise your estimate?

Participants may have a variety of ideas. Allow enough time for them to share their thoughts.

Collect the Data

Participants open their boxes and count the raisins.

Analyze the Data

As participants finish their counts, they report their data. You may want to begin by asking who thinks they have the smallest number of raisins. Once this has been determined, do a similar check on the largest number of raisins. This permits you to establish the range—the upper and lower boundaries of the data set—so that you can make a line plot of the data.

Organize the raisin data on a line plot large enough for everyone to see (on an overhead transparency or chart paper) by having each person tell you the number of raisins in his or her box. A sample line plot is shown below.

Raisins in Our Boxes

```
            X              X
            X              X
            X   X          X
         X  X  X  X        X        X
         X  X  X  X     X  X        X
         X  X  X  X  X  X  X  X  X     X
        ─────────────────────────────────────
        26 27 28 29 30 31 32 33 34 35 36 37 38 39 40 41 42
                         Number of raisins
```

What are some of the things we can say about the data? Suppose someone asked you, "About how many raisins are in a box?" What could you say?

Help participants focus on the five components that can be used to describe the shape of the data (see Transparency 6.a).

- Is the shape of the distribution symmetrical or skewed? What does it mean to be symmetrical? What does it mean to be skewed?

- Are there single or multiple peaks? What does this tell you about the data?

- What is the typical number of raisins in a box? (A variety of responses are possible, including computing one or more measures of center.)

- How spread around the center of the data are the rest of the data? What does this tell you about the data?

- Are there any unusual deviations from the patterns you found in the data, including gaps and outliers?

Interpret the Results

What do you expect to happen to the shape if we collect more data? If we opened five more boxes of raisins, what is your best guess of how many raisins would be in each box, based on the data we already have?

Teacher Notes

From this display, we can see that approximately half of the boxes have 28–32 raisins and half of the boxes have 34–40 raisins. Although the range is 28–40 raisins, a large clump of data falls in the interval of 28–32 raisins, with a smaller clump in the interval of 34–36 raisins. An unusual value, or outlier, occurs at 40 raisins. The three boxes at 38 raisins may also be outliers. There is a gap between 32 and 34 raisins.

Allow participants to work on this question for a few minutes in small groups, then report their theories to the class. Encourage them to ask each other questions, and expect them to explain their ideas. At the end of the session, allow participants to eat the raisins.

Developing the Activity: Part 2

Participants will determine the years of the pennies they have with them (or use the data provided), record and organize the results, and describe the shape of the data distribution.

Pose the Question

Have participants respond to this question:

What are the dates on the pennies you have with you now?

Collect the Data

In this portion of the activity, it is assumed that participants have not collected their own data, though they may want to do so. Show the two sample distributions of data (Distributions A and B on Transparencies 6.c and 6.d and on Handout 6.1). If participants decide to collect their own data, display it on a transparency or chart paper using a line plot.

Analyze the Data

Have participants consider Distribution A. Discuss the shape of the data. These data were collected in Fall, 1994.

What can we say about the shape of this distribution? What does it tell us about the dates of pennies in this collection?

Discuss the fact that this is an example of a distribution that is skewed to the left.

Clearly, most of the data is clumped in the 1980s and, particularly, at 1994. What's typical about these data? What can we say about the spread? Do any patterns emerge?

Now have participants consider Distribution B.

Is this distribution similar to Distribution A? What does it tell us about the shape of the data?

Interpret the Results

Participants use the distributions of data and results of their discussion to respond to this question:

What can we say about the circulation of pennies?

Participants often suggest that collecting pennies is a popular hobby, which may be one reason why pennies with earlier dates are not often in circulation. This speculation lends itself to a good social studies project about the manufacturing and distribution of currency.

Participants are likely to discover that the two sample distributions are quite similar.

We suggest that you collect data on dates of pennies the participants have, describe the shape of the distribution, and compare those data with the two graphs showing 1994 data.

Summary

Describing and interpreting data are skills that are developed over time. Looking for patterns or trends in a data set moves participants beyond focusing on individual numbers and isolated bits of information.

Descriptive words provide visual images. Encourage the use of descriptors such as *clumps, clusters, bumps, gaps, holes, spread out, bunched together, shaped like a bell,* and *shaped like a ski jump*.

Discussions with respect to the shape of the data focus on two aspects: (1) identifying the special features of the shape of the data, and (2) deciding how to interpret this information in terms of theories and experiences that might account for the shape of the data.

Overall Shape of the Data

- Is the distribution symmetrical or skewed? What does it mean to be symmetrical? What does it mean to be skewed?

- Are there single or multiple peaks? What does this tell you about the data?

- What can we say is typical about these data?

- How spread around the center of these data are the rest of the data? What does this tell you about the data?

- Are there any unusual deviations from the patterns you found in these data, including gaps and outliers?

Distribution A

Dates of Pennies

```
                        x x x x x x x x x x x x x x x
                                          x x x x
                                        x x x x x x
                                          x x x x
                                            x x
                                        x x x x x x
                                          x x x x
                                            x x
                                        x x x x x
                                      x x x x x x x
                                      x x x x x x
                                        x x x x x
                                            x x
                                            x x
                                            x x
                  x x x x x
                    x x x x

            x

x                                                                  
x x
x
66 67 68 69 70 71 72 73 74 75 76 77 78 79 80 81 82 83 84 85 86 87 88 89 90 91 92 93 94
```

Year (19--)

Distribution B

Dates of Pennies

```
                        x x x x x x x x x x x x x x x x x x x x
                                              x x x x x
                                        x x x x x x x x
                                                x x
                                            x x x x x
                                                x x
                                              x x x
                                              x x x
                                              x x x
                                              x x x
                                          x x x x x
                                              x x
                                      x
            x x
          x
x
61 62 63 64 65 66 67 68 69 70 71 72 73 74 75 76 77 78 79 80 81 82 83 84 85 86 87 88 89 90 91 92 93 94
```

Year (19--)

Distribution A

Dates of Pennies

```
                        x x x x x x x x x x x x x x
                                        x x x x
                                      x x x x x x
                                        x x x x
                                         x x
                                      x x x x x x
                                        x x x x
                                         x x
                                       x x x x x
                                      x x x x x x
                                      x x x x x x
                                       x x x x x
                                         x x
                                         x x
                                         x x
                                x x x x
                               x x x x
                        x
x x
──────────────────────────────────────────────────────
66 67 68 69 70 71 72 73 74 75 76 77 78 79 80 81 82 83 84 85 86 87 88 89 90 91 92 93 94
```

Year (19--)

HANDOUT 6.1

HANDOUT 6.1

Distribution B

Dates of Pennies

```
                              x x x x x x x x x x x x x x x x x x x x x x
                                                    x x x x x x
                                          x x x x x x x x x x
                                                      x x x
                                            x x x x x
                                              x x x
                                              x x x x
                                                x x x
                                              x x x x
                                          x x x x x
                                  x x
                              x
       x
 x
 ─────────────────────────────────────────────────────────────────────
 61 62 63 64 65 66 67 68 69 70 71 72 73 74 75 76 77 78 79 80 81 82 83 84 85 86 87 88 89 90 91 92 93 94
```

Year (19--)

INVESTIGATION 7
Shape of the Data: Line Plots to Bar Graphs

Overview

Participants explore the theoretical differences and similarities between graphically representing data using line plots and using bar graphs. They first focus on tallying frequencies and then work to construct bar graphs using appropriate labels on the vertical (y) axis and horizontal (x) axis.

Assumptions

Before working with bar graphs, participants have used line plots and considered ways to describe the shape of the data.

Goals

Participants explore the grouping of data through graphical representations. In particular, they

- make theoretical distinctions between line plots and bar graphs, identifying the role played by the y-axis

- work with the concept of grouping data by using tallies to determine the frequency of occurrence of each distinct data value

References

Curcio, F. R. *Developing Graph Comprehension*. Reston, Virginia: National Council of Teachers of Mathematics, 1989.

Fey, J., W. Fitzgerald, S. Friel, G. Lappan, and E. Phillips. *Data About Us*. Palo Alto, California: Dale Seymour Publications. Forthcoming.

Materials

Transparency marker, chart paper, calculators

Transparencies 7.a through 7.c

Handouts 7.1, 7.2

Developing the Activity

Project Transparency 7.a, which shows a line plot and a table of data.

Pose the Question and Collect the Data

Have participants explain what data are being shown and what the line plot tells them about the data.

Focus the discussion on the table.

How is this table of data related to the line plot?

What does it mean when we say this is a "frequency table showing data about lengths of names"?

Participants should see that the table provides a tally of the counts of lengths of names and a summary of the data that was used to make the line plot.

Using Transparencies 7.a and 7.b, explore how the line plot is similar to and different from the bar graph. One way to demonstrate the relationship between the line plot and the bar graph is to draw bars over the Xs on the line plot using a transparency marker. You can also color the bars to make the point that we can no longer distinguish each data item and therefore need a way to show the count, or frequency, of each occurrence of the distinct values. This is why we must introduce the vertical axis.

Discuss how data from the frequency table may be used to construct the bar graph.

Analyze the Data

Provide each participant with a copy of Handout 7.1, the problem sheet for this investigation. There are two problems to consider; for each, participants are asked to make a frequency table and a bar graph and to write questions about the data displayed in the graphs.

The first problem involves scaling the x-axis, beginning at 0. For the second problem, participants may choose to scale the x-axis beginning at some other number (for example, 26). The convention for indicating a break in the scale (for instance, from 0 to 26) is to show two slanted lines (//) interrupting the x-axis.

When computer software is used to create graphs, this convention is generally not followed.

With bar graphs, the convention is to make bars that do not touch one another.

Using Transparency 7.c, spend some time discussing the three components to graph comprehension (Curcio 1989).

- *Reading the data* involves "lifting" the information from the page to answer explicit questions for which the answers are clearly displayed in the graph. For example, this question involves reading the data: How many names have 12 letters?

- *Reading between the data* involves the interpretation and integration of information presented in a graph. This includes making comparisons (for example, greater than, greatest, tallest, smallest) and the use of other mathematical concepts and skills (for instance, addition, subtraction, multiplication, division). This question involves reading between the data: How many names have more than 12 letters?

- *Reading beyond the data* involves extending, predicting, or inferring from data to answer implicit questions. The reader gives an answer that is not literally present in the graph to a question that is at least related to the graph. For example, this question involves reading beyond the data: If a new student joined the class, what do you predict will be the number of letters in that student's name? Explain your reasoning.

Interpret the Results

Have participants review the questions they wrote with respect to the two graphs on Handout 7.1 and classify them based on the three components of graph comprehension: *reading the data, reading between the data,* and *reading beyond the data.* You may want to make three chart pages, one for each component, and have participants record their questions so you have results for the entire group. You can shorten the process by having participants record only questions that don't duplicate those already recorded.

Participants may want to explore developing other questions that are related to one or more of the three question classifications. Generally, *reading the data* and *reading between the data* questions are the most common categories of questions asked; asking and responding to *reading beyond the data* questions is often more challenging.

Summary

It is not obvious to students (or sometimes adults!) that line plots, frequency tables, and bar graphs can be considered companion representations. Spending time properly highlighting the distinctions among representations is important.

The three classifications that relate to graph comprehension need to be considered throughout the remaining investigations. Using the vocabulary of *reading the data, reading between the data,* and *reading beyond the data* can help participants as they consider the level of questioning they are using.

Letters in First and Last Names in a Grade 6 Class

```
                            X
                  X         X
                  X         X
        X         X  X  X   X
        X      X  X  X  X   X
        X      X  X  X  X   X      X      X
  ──────────────────────────────────────────────
   7    8   9  10 11 12 13  14  15  16  17
```

Number of letters

Number of Letters in Names

Letters in name	Frequency
8 letters	3
9 letters	0
10 letters	2
11 letters	3
12 letters	5
13 letters	3
14 letters	6
15 letters	1
16 letters	1

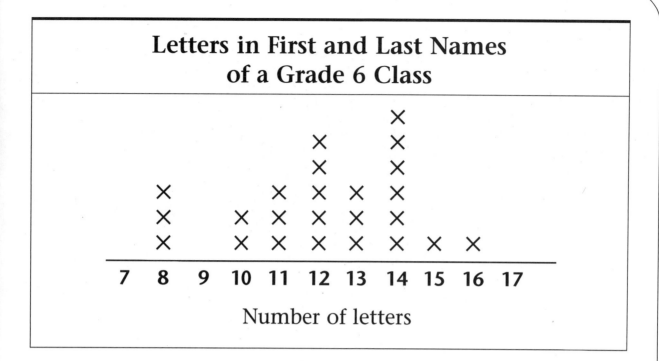

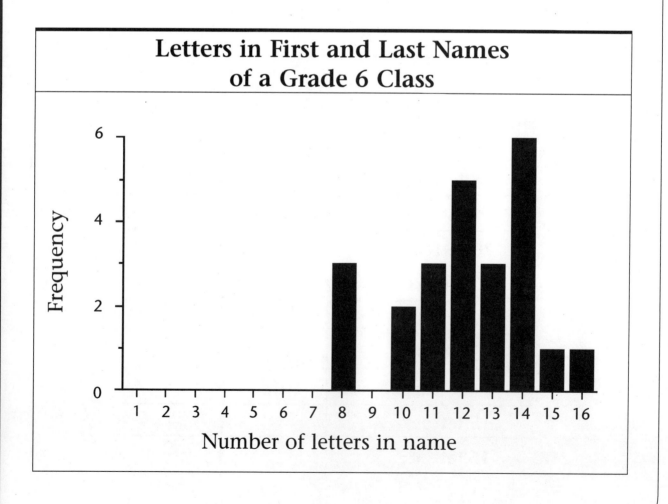

Three Components to Graph Comprehension

Reading the data

Reading between the data

Reading beyond the data

Reference

Curcio, F. R. *Developing Graph Comprehension.* Reston, Virginia: National Council of Teachers of Mathematics, 1989.

Line Plots to Bar Graphs Problem Sheet

1. A grade 4 class investigated the question of how many pets students in the class had. They had a discussion about what "counted" as a pet and then gathered their data. Here are their results, shown in a line plot.

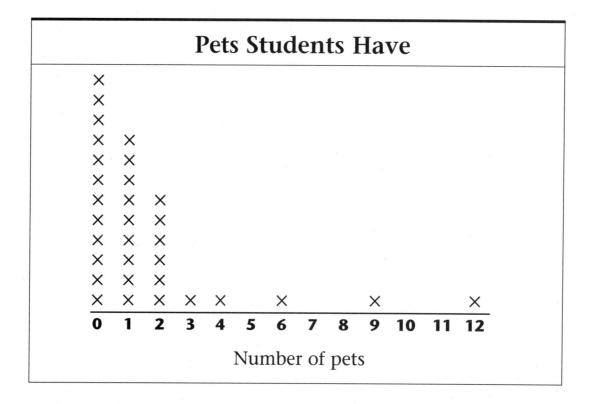

   a. Make a frequency table and a bar graph to show the information that is displayed in the line plot.

   b. Write several questions you might ask students with respect to these data.

2. One grade 6 class investigated the question of how many raisins are in a box of raisins. They each had a half-ounce box of raisins. They estimated and then counted the number of raisins.

   Here is their data.

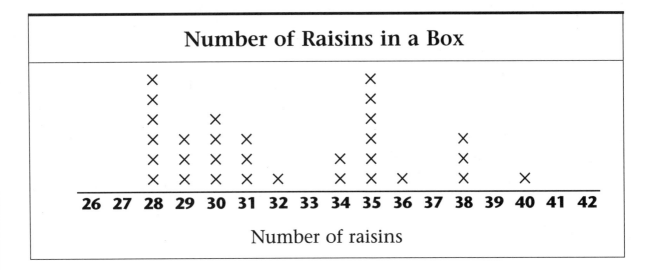

a. Make a frequency table and a bar graph to show the information that is displayed in the line plot.

b. Write several questions you might ask students with respect to these data.

Three Components to Graph Comprehension

Reading the data involves "lifting" the information from the printed page to answer explicit questions for which the answers are clearly displayed in the graph. An example of a question that addresses reading the data is "How many people had 35 raisins in their box?"

Reading between the data includes the interpretation and integration of information presented in a graph. This includes making comparisons (for example, greater than, greatest, tallest, smallest) and the use of other mathematical concepts and skills (for instance, addition, subtraction, multiplication, division). Implicit questions involve at least one step of logical or pragmatic inferring to get from the question to the response. Both question and response are derived from the graph. Examples of questions that address reading between the data are "How many people had more than 34 raisins in their box?" and "How many more people had 28 raisins in their box than had 29 raisins in their box?"

Reading beyond the data involves extending, predicting, or inferring from data to answer implicit questions. The reader gives an answer that is not literally present in the graph to a question that is at least related to the graph. Examples of questions that address reading beyond the data are "What is the typical number of raisins in a half-ounce box of raisins? If you open another box of raisins, how many might you expect to find?"

Reference

Curcio, F. R. *Developing Graph Comprehension.* Reston, Virginia: National Council of Teachers of Mathematics, 1989.

INVESTIGATION 8

Giant Steps, Baby Steps

Overview

The previous session was spent on one of the first steps in collecting data: identifying attributes or properties of objects and producing data by counting. This session will focus on producing data by a different type of counting, which we refer to as *measurement*. Two important aspects of measuring properties—precision or accuracy, and scales of measurement—will be addressed in other investigations.

Assumptions

Participants are able to state the four basic steps of the process of data analysis. They should understand that people, objects, concepts, and constructs are entities with characteristics, attributes, or properties that give rise to questions that lead to the collection of data. When collecting data, characteristics, attributes, and properties can either be classified (categorized) or counted (measured). Once the question is classified as answerable or unanswerable, it is clarified, refined, and revised as necessary. During this process, the investigator will identify a property or variable to investigate or measure.

Goals

Participants explore the process of data investigation. In particular, they

- identify strategies for collecting data

- understand ways to collect data appropriately

- collect real data through experiments

- recognize how errors may present themselves in the course of the data-collection process

- identify misrepresentations of data

- clarify, refine, and revise the questions being asked

- understand ways to record data appropriately

Materials

Stick-on notes, calculators

Transparencies 8.a, 8.b

- represent data using line plots
- describe landmarks and features of data

References

Corwin, R. B., and S. J. Russell. *Measuring: From Paces to Feet.* Palo Alto, California: Dale Seymour Publications, 1990.

Moore, D. S. *Statistics: Concepts and Controversies*, 3rd ed. New York, New York: W. H. Freeman and Co., 1991.

Developing the Activity

Begin this session with an overview of the topic of measurement. Point out how this topic of measurement connects to the previous session on attributes.

Noticing attributes or characteristics of objects often leads to posing questions about those objects. Once the question about the object has been formulated, the next step is to collect data on the characteristics or attributes of interest.

The type of data we gather to help us answer our question will vary. Sometimes we "measure" the attribute, or variable, by first sorting or classifying the objects into categories such as "with antennae" and "without antennae." Sometimes we measure by counting objects, such as the number of raisins in a box. When we count objects like these, we use a discrete scale of measure. A discrete number scale has a distinct number of values between any two given values. For example, shoe sizes are a discrete scale of measure, because between a size 9 shoe and a size 7 shoe, there are only three possible values ($7\frac{1}{2}$, 8, and $8\frac{1}{2}$).

When we measure a variable such as length, we are measuring with a continuous scale of measurement, because, theoretically, between any two given values there are an infinite number of other values. For example, between 1 inch and 2 inches are 1.25 inches, 1.33 inches, 1.6 inches, and so on.

It is critical to the success of the data-analysis process that the data are gathered accurately and consistently. What can happen if the data are not gathered accurately and precisely? [Wrong conclusions can be drawn.]

In the investigation that follows, we will consider some of the mechanical and conceptual issues involved in measuring variables.

Pose the Question

What is the length of this room?

Solicit a variety of types of unit measures (for example, 32 feet, 9 meters, 10 yards). Record estimates on the board or overhead projector. Underline the unit of measure in each estimate.

Is the length of this room 9, 10, or 32?

Length depends upon the unit used to measure.

Let's refine the question: How may "giant steps" long is this room?

Solicit estimates, and record them in order from least to greatest.

What is the range of this data?

Accept the difference between greatest and least, as well as the two numbers at the end points, such as 13–22.

Why is there such a range in our estimates?

The number of giant steps depends upon the size of the stride of the person who makes the stride. Also, some people may be better at estimating than others.

What types of numbers would we have for estimates if we had the basketball team respond to this question? What types of estimates would we have if we asked a roomful of grade 2 students?

The people with the larger stride, or unit of measure, would give smaller estimates than the people with shorter strides.

We say there is an inverse relationship between the size of the unit and the number of units required to measure a distance. Smaller units yield larger numbers.

Because the size of a giant step varies, we say it is a **nonstandard** *measure. Elementary-grades children begin their measuring experiences with nonstandard units. Give some examples of nonstandard measures your students have used that reflect length or distance.*

Examples might include measuring with paper clips, interlocking cubes, strides, or paces.

Collect the Data

Let's return to our question: How long is this room? Take turns with a partner as you do the following:

- Estimate the length of the room in your giant step. Record your estimates on scrap paper. (Note: "Length of the room" must be defined and agreed upon by the group.)

- Pace the length of the room in giant steps, and record your measurements on stick-on notes.

- Post the stick-on notes on the board.

As participants collect data, circulate to respond to questions. Teachers may ask questions such as, "How big should your giant step be—a lunge? A normally large step?" This is a good question, and participants should note that it is important to raise with the entire group when they begin analyzing the data. They may also note that there may be variation in the number of giant steps each person finds for the length. In other words, multiple trials might produce more reliable data. If teachers wish to repeat their measurements, this is fine, but not necessary. The point about variation should be reiterated during the interpretation stage of the investigation.

Analyze the Data

Once the data are posted, ask the following questions:

How can we describe the data in general terms? What is the shape of the data?

There is no shape yet. The data must be organized first.

How could we organize the data?

The data can be organized from smallest to largest in a (properly labeled) line plot.

Is the data symmetrical or skewed? Is there a single peak? Are there multiple peaks? Are there any holes, or gaps? Are there any unusual values, or outliers?

How much variation is there? What is the range?

Accept but do not encourage descriptions of the data using computed statistics such as mean, median, and mode.

Interpret the Results

What do smaller/larger values represent?

Smaller/larger values represent taller/shorter people.

How would you answer the question, How long is this room?

Answers might employ justification, mean, median, mode, or perhaps range.

How could you apply this information?

If you know that the room is *n* giant steps long, you could estimate the length of other rooms.

How might we refine or extend this question?

We could ask, What is the relationship between the height and length of the room in paces?

Summary

Project Transparency 8.a, and summarize the key points of this investigation.

1. There is a need to define units of measure precisely before gathering data.

2. A large range in a set of data may mean that the units of measure were not precisely defined or that nonstandard units were used.

3. When analyzing data, it is important to consider the units used to gather the data.

4. Sometimes statisticians use small units to get big numbers and vice versa, depending on the impact they want to have.

5. A variable such as length or distance can be described in different ways, depending on the units or scales used to measure.

End by discussing Transparency 8.b, and allow participants to share some of their own rules of thumb.

Sample Questions

What is a "giant step"? Is it a lunge?

How would the data change or look if we had used "baby steps" to collect the data?

Which sounds bigger, 20 giant steps or 40 baby steps?

Which sounds longer, 9 weeks or 63 days?

Giant Steps, Baby Steps: Summary

1. There is a need to define units of measure precisely before gathering data.

   What is a "giant step"? Is it a lunge?

2. A large range in a set of data may mean that the units of measure were not precisely defined or that nonstandard units were used.

3. When analyzing data, it is important to consider the units used to gather the data.

   How would the data change or look if we had used "baby steps" to collect the data?

4. Sometimes statisticians use small units to get big numbers and vice versa, depending on the impact they want to have.

   Which sounds bigger, 20 giant steps or 40 baby steps?

   Which sounds longer, 9 weeks or 63 days?

5. A variable such as length or distance can be described in different ways, depending on the units or scales used to measure.

Rules of Thumb

Americans stand just far enough apart when talking that, with arms extended, they can insert their thumbs into each other's ears.

At arm's length, the width of a paper matchstick covers the space 5 yards wide 1000 yards away.

The smaller the bird, the closer it will allow you to approach.

Reference

Parker, T. *Rules of Thumb 2.* Boston, Massachusetts: Houghton Mifflin, 1987; pp. 80, 11, 60.

INVESTIGATION 9
What Is the Typical Foot Length of Our Group?

Overview

This session focuses on producing data by measurement. Two important aspects of measuring properties—(1) precision or accuracy, and (2) scales of measurement—will be explored. Participants make estimates of their foot lengths and then use footsticks to measure.

Assumptions

Participants are able to determine whether or not a question is answerable; to clarify, refine, and revise a question as necessary; and to identify a property or variable to investigate or measure.

Goals

Participants explore the concept of measurement within the context of the process of statistical investigation. In particular, they

- identify strategies for collecting data

- understand ways to collect data appropriately

- collect real data through experiments

- recognize how errors may present themselves in the course of the data-collection process

- identify misrepresentations of data

- clarify, refine, and revise the question being asked

- understand ways to record data appropriately

- represent data using line plots

- describe landmarks and features of data

Materials

Footsticks (1 per team), inchsticks (1 per team), stick-on notes, pens, calculators

Transparency 9.a

See "The King's Foot" in *Measuring: From Paces to Feet* (Russell and Corwin 1990), beginning on pp. 40–41.

References

Russell, S. J., and R. Corwin. *Measuring: From Paces to Feet.* Palo Alto, California: Dale Seymour Publications, 1990.

Blocksma, Mary. *Reading the Numbers.* Bergenfield, New Jersey: Penguin Books, 1989.

Developing the Activity

In collecting data to answer a question, we have found that you can classify and count items or you can measure items. Actually, measuring can be thought of as a type of counting, too. For example, when we measured the length of the room, we counted the number of giant steps it took us to cross the room. If we say that the typical giant step is this long [hold hands approximately 3 feet apart], *we could cut a piece of adding-machine tape equal to the length of the typical giant step and see how many times we would have to lay it end to end to reach from one side of the room to the other.*

To set the stage for our next investigation, let's hear about a king who measured things much the same way we did in our last investigation.

Read "The King's Foot" in *Measuring: From Paces to Feet*, pp. 40–41.

Have any of you ever had an experience with measurement like the one in this story?

Allow time for discussion.

Why was the new stall too small at first? How did the carpenter solve the problem? Why did she call the sticks she made "rulers"?

In this investigation, we will collect data using standard and nonstandard units of measure and explore some of the issues related to collecting data by measuring with standard units.

Pose the Question

What is the typical foot length of people in our group?

Record estimates on the board or overhead. Accept a variety of units, such as 10 inches and size 7. Allow time for discussion.

From our estimates, it seems as if the answer to our question will depend, among other things, upon the units we use. To begin with, we will use a footstick as our tool.

Put a chart like this on the board:

Shorter Than Footstick	Equal to Footstick	Greater Than Footstick

Collect the Data

Distribute footsticks, and ask participants to compare their own feet with the rulers and to place a tally mark on the board in the appropriate column.

Analyze the Data

What can you say about the data?

Most feet will be shorter or less than the footstick.

How much variation in foot length is shown is this chart?

We can't tell much about variation from this display.

Does that mean everyone's feet are the same length?

Interpret the Results

Why is there so little variation in our foot lengths according to the data?

Because the tool was not very precise, and because the chart only categorizes data in three categories, which doesn't tell us very much.

This exercise produced *categorical* or *nominal* data. This kind of data provides only general information. More discussion about types of data will come later.

How would we answer the initial question (What is the typical foot length of people in our group?) based on the data in our chart? How could we find out more precisely the typical length of our feet?

A more precise measuring tool could be used.

Re-Collect the Data

Distribute the inchsticks.

Here is a tool that can help us measure more precisely. How is this tool different from the footsticks?

The inchsticks are divided into smaller units.

Let's repeat our earlier data-collection process, this time using the inchsticks. Your task is to measure your partner's foot length with an inchstick. Record your data individually on a stick-on note, and put it on the board.

Questions will arise regarding standardization of the measuring protocol, such as, Should we measure with shoes on or off? Sitting down or standing up? Right foot or left foot? Do we round up or express our data as parts of whole units? These are good questions and should be brought up again during the data-analysis stage.

Analyze the Data

How can we describe this data in general terms? What is the shape of this data?

There is no shape yet. The data must be organized first.

How could we organize the data?

Data could be organized from smaller to larger units. At this point, you may see that some measurements are in whole units and others are in mixed numbers. If so, ask about this.

What might account for the different types of numbers we see here?

Some measures are to the nearest whole unit.

Agreeing on measurement and recording procedures prior to data collection is very important, especially when you have multiple researchers involved in a project. What other factors might account for the variations we see?

Other factors might include measuring with shoes off versus on, sitting versus standing, right foot versus left foot. At this point, discuss and agree on protocols to be used, and then remeasure.

Post the new, standard measurements on a different part of the board. Select one or two teachers to organize the data as it is brought to the board.

Let's look at the general shape of these data. Are they symmetrical or skewed? Is there a single peak? Are there multiple peaks? Are there any gaps or holes? Any unusual values or outliers? How much variation is there? What is the range?

Accept but do not encourage descriptions of the data with computed statistics such as mean, median, and mode.

How can we find the middle-size foot from this graph?

Accept responses, being careful to note whether "middle" is determined by the middle data point or the middle of the range on the horizontal axis. For example:

What is the middle-size foot (in inches) shown on this graph?

```
                        X
                        X    X    X
              X    X    X    X    X
     ─────────────────────────────────────
     3   4   5   6   7   8   9   10   11
```

The middle-size foot is 8 inches in length.

How does that compare with the middle-size foot on this graph?

```
                   X
                   X    X    X
         X    X    X    X    X
         ─────────────────────────
         6    7    8    9    10
```

It is (or should be) the same.

Note that "middle" refers to the middle data point, not the middle number of the range shown on the graph.

Interpret the Results

One of the last steps in the process of data analysis is to ask whether there are further questions or extensions of the original question that now warrant analysis.

Can you think of other questions that might arise as a result of this investigation?

Relationships between different body measures will be explored further in Investigations 41–44.

Sample Questions

Is there a relationship between foot length and shoe size?

Will the distribution of shoe sizes be the same as the distribution of foot lengths?

Summary

Display Transparency 9.a.

How precise your measurements are depends on the tool that you use. The "right" tool is needed for each task.

Using different tools and having different people taking measurements may produce different data values for the same task. When we use conventional tools and standard measures, we improve reliability and precision and reduce variability.

Typical Foot Length: Summary

1. Precision of measurements is affected by the tool used.

   Finer divisions in a measuring tool allow you to be more precise.

2. You need the "right" tool for the task.

   For example, the smaller the tool you use to measure the length of the room, the greater the chance you will make a measurement error, as you will have to count more of a smaller unit, such as inches, than of a larger unit, such as yards.

3. Different tools and different measurers may produce different data.

   When collecting data, there is a need to standardize measuring tools and data-collection protocols.

4. Conventional (versus nonstandard) tools improve reliability and precision and reduce variability in data.

INVESTIGATION 10
Children and Measurement: A Minilecture

This minilecture is based on what research says about children's understanding of measurement.

Transparency 10.a

Begin by showing this title transparency.

Transparency 10.b

Group teachers into four groups. Give each group a blank transparency and marker, and ask them to list all the measurements they came across from the time they woke this morning until now. (For example: mixing a 6-ounce can of orange juice with 3 cans of water; determining the caloric content of 8 ounces of breakfast cereal if 3 ounces contain 90 calories; driving 55 miles per hour for 45 minutes; filling an 8-gallon gas tank at $1.29 per gallon; traveling 120 miles if the car gets 30 miles per gallon.) Display each group's transparency for participants to discuss.

Measurement is a vital component of our everyday lives, and we express measurements in a variety of ways.

Transparency 10.c

What do you think the first statement means? Length can be measured in such standard units as feet, inches, yards, meters, miles, and kilometers.

Students must have an idea of what length each unit expresses and how to determine which unit is the most appropriate in a given situation. If you were to measure the distance from Raleigh, North Carolina to Rocky Mount, North Carolina, what unit of measurement would be most appropriate? Why?

Possible answers: Miles, kilometers. These are large units and would result in fewer errors in counting than would smaller units.

Suppose we were making orange juice from concentrate and the directions read, "mix one can of frozen concentrate with three cans

Materials

Blank transparency sheets, transparency markers

Transparencies 10.a–10.m

of water." What is the unit of measurement? Is it a standard unit of measurement?

Answer: The unit of measurement is an orange juice can. This is not a standard unit of measure.

Review the lists generated earlier, and look for any nonstandard units of measurement.

Can you think of some examples in which estimating size would be more appropriate than determining a precise measurement?

Possible answer: Estimating the cost of items while grocery shopping might be more appropriate than figuring precisely.

Can you think of some examples in which a precise measurement would be more appropriate than an estimate?

Possible answer: Carpentry requires precise measurements.

Are the items on our lists precise measurements or estimates?

Transparency 10.d

A *comparison* doesn't tell us anything about the actual measurement. If we do not know today's temperature and someone says it is cooler today than yesterday, we can't know whether it is cooler by 1 degree or 30 degrees. In some cases, just knowing it is cooler is sufficient.

Sometimes using an *average* is more appropriate than using multiple, precise units of measure. Instead of listing the number of miles we drive each day for a week, we can summarize with an average of, for example, 30 miles per day. Returning to our groups' lists, let's see if anyone gave an example of a measurement that represents an average.

Rates imply fractions. For example, miles per gallon = miles/gallon; kilometers per hour = kilometers/hour. Were any of our measurements rates?

Transparency 10.e

Children develop spatial concepts on two levels. *Perception* is the ability to perceive an object through the senses. *Representation* is the ability to imagine an object in the mind. For representation to be successful, children must first experience perception. As teachers, we can help by exposing them to both perception and representation. Perception experiences are more important to younger children because they will later help them with representation.

Students should practice visualizing and moving objects in their minds. For example, to introduce a rectangle, we let children first physically manipulate it. Then, using transparencies of geometric shapes on the overhead, we might discuss what happens when the rectangle is rotated, flipped, and reflected. Later we may take the object away and have the children draw it in different situations.

Studies have shown that boys are often more skilled at spatial visualization than girls. This may be due to the fact that boys play with blocks and spatial objects more frequently than girls. We need to give both girls and boys plenty of experiences with spatial manipulation.

Spatial manipulation is an important part of measuring. When children learn about concepts such as area and perimeter, they will likely be working with spatial objects such as rectangles, squares, and triangles.

Transparency 10.f

The more active a student is in working with spatial objects, learning concepts such as area, volume, measuring objects, and relating measurement to the concepts learned, the more likely the student will retain the information.

Students need to be involved on many levels. First, they might examine a rectangle (perceive it and make a representation of it). Second, they might learn to find the area of the rectangle. To do this they must have a concept of area and an understanding of the link between finding area and measuring. Third, they should reflect on what they have done. This reflection period may occur in a discussion format or by having students write about their experience.

Here are two approaches to teaching perimeter.

Approach 1

Meela is teaching perimeter to her students. She writes the definition of perimeter from the textbook on the board: $P = S + S + S + S = 4S$. She tells students that *perimeter* is the distance around an object. She then puts examples on the board from the textbook and demonstrates how to calculate perimeter for different objects. Students are given practice problems computing perimeters to do at their desks. Meela teaches area in a similar fashion and wonders why her students get area and perimeter confused on a test.

In this example, students are not building concepts, testing hypotheses, or challenging new ideas. They are simply following an algorithm that says "add the lengths of each side of the object."

Approach 2

Laura says to her students, "Your dad tells you to take the dog for a walk around the *perimeter* of the yard. What do you think he wants you to do?" Laura helps guide students' answers and hypotheses in the right direction. After this introduction of the word in context, she writes PERIMETER on the board. She asks some students to walk the perimeter of the classroom. After moving the desks out of the way, she makes a square on the floor using string. Students walk the perimeter of the square.

Laura continues the lesson with different shapes. After demonstrating a rectangle, she asks students how they could use string to create a triangle with the same perimeter as the rectangle. Finally she brings numerical values into play, and the students discover for themselves how to compute an object's perimeter. They also develop such formulas as: Perimeter of a rectangle = 2(length) + 2(width). This is part of the discovery and concept-building process.

By starting with the concept and working toward discovering the algorithm, students will retain an understanding of the concept, an essential outcome.

Transparency 10.g

Many of us were, unfortunately, simply told to memorize the rules of mathematics. Later we would forget the rules and wonder why we couldn't understand the concept that related to the formula or rule.

Today's students are being taught differently. Memorization is de-emphasized. Instead, it is recommended that teachers teach concepts. Once students understand a concept, they can derive the rule for themselves. Students will have a much better grasp of mathematics in that they will be able to apply what they are learning to new situations and real-world problems.

Transparency 10.h

For students to successfully master measurement—and concepts such as area and volume that are related to measurement—they must understand conservation, transitivity, and the meaning and the importance of a unit.

Transparency 10.i

Students should be able to conserve length, area, and volume. Let's look at how we could determine whether our students can conserve length, area, and volume.

Length

This can be acted out as if you were testing children.

Show students the two straws aligned next to each other. Ask whether the straws are the same length. Now move one straw slightly ahead of the other. Students who can conserve length will still agree that the straws are the same length. Those who cannot conserve length will say that one straw is now longer than the other.

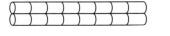 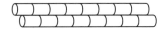

Area

This can be acted out as if you were testing children.

Cut one sheet of construction paper in half, and then cut one of the half sheets into nine equal squares. Cover half of the full sheet of paper with the nine squares. Ask the children to observe the amount of red area (assuming your full sheet is red) they can see.

Randomly scatter the pieces on the paper. Ask the children to compare the red area they saw before to the red area they can see now. Children who can conserve area will say that the two areas are equal. Children who cannot conserve area will tell you there was more area the first time than there is now.

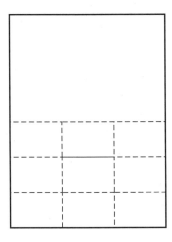 

Volume

This can be acted out as if you were testing children.

Place three containers (plastic food containers work well) in front of the students. One container should be rectangular, and the other two should be cylindrical and the same size.

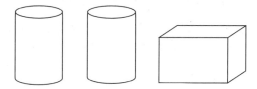

Fill the two round containers about three-fourths full with water so that they are at the same level. Then, ask the children, "Does this container have more, less, or the same amount of water as this other container?" Children will probably answer that they have the same amount. Pour water from one of the round containers into the rectangular container. Ask the children, "Does the rectangular container have more, less, or the same amount of water as the round container?" Children who can conserve volume will know that even though the water lines differ, the amount of water is still the same. Children who cannot conserve volume will immediately tell you that the container with the higher water line has more water. Finish by pouring the water from the rectangular container back into the round container and asking the children the question again. They should respond that the containers have an equal amount of water.

Piaget said that conservation is a developmental task and that children should be able to conserve around the age of seven. He also believed that once a child could conserve one attribute, such as length, that child should be able to conserve all attributes. Subsequent research has shown this isn't the case. Children's cognitive development has moments of being between stages, so they may be able to conserve length but not yet be able to conserve volume.

Transparency 10.j

Transitivity of measurement is similar to the transitive property in mathematics. The transitivity property tells us that if $a < b$ and $b < c$ then $a < c$ must be true. For example, if $2 < 5$ and $5 < 7$ then $2 < 7$.

This idea can be presented to children in a measuring context. If you have three pencils and the first pencil is shorter than the second pencil, and the second pencil is shorter than the third pencil, then the first pencil must be shorter than the third pencil. Students can discover this property for themselves by doing an activity in which they repeatedly prove this fact until they notice a pattern.

Transparency 10.k

Students must learn to determine which attribute is being measured. This is most effectively done by allowing students to say what they think is best and then justify their choice. Students can explore the difference between one-, two-, and three-dimensional space by examining length, area, and volume.

Transparency 10.l

After a lot of experience, most adults have learned how to focus on several dimensions, the number of units, and the size of the unit concurrently. Students need experience to gain this ability.

Focusing on several dimensions is a complex task, and students will need your help. The importance of assigning or identifying a unit should be stressed. When we say 5, it has no real-world meaning if it is not attached to a unit of measure: 5 inches is very different from 5 miles. Students have just begun to master number concepts. Adding and subtracting have been recently learned, so the number part of the answer will play the strongest part. In all problems, the unit of measurement must be stressed, and students must understand the significance that units of measure attach to numbers.

Transparency 10.m

The process of learning to measure can be divided into five sequential steps designed to span grades K–5.

First, students must be able to identify the property to be measured. Are we measuring length, distance, volume, area, weight, or time?

Second, students must be able to make comparisons. For example, they might be given the task of comparing the heights of two students. Students might discuss different methods of doing this, such as standing back-to-back and observing the difference, or actually measuring the height of each student and comparing the measurements. Students can discuss which method they think is better, more accurate, easier, and so on.

Third, students must be able to establish an appropriate unit for measurement. Both standard and nonstandard measurements should be explored and discussed. Students can measure the height of a classmate by using many different units of measurement—for example, inches, centimeters, feet, shoe boxes—then discuss how precise each measurement is and which measurement is preferred.

The fourth step in the process of learning to measure involves experiences with standard units of measurement. Students can further discuss which unit is appropriate. Students should practice estimating with various units. Students in the upper grades can also look at error measurement and precision in measuring.

Fifth, work with students to create formulas to help them count units. Students will naturally develop shortcuts to counting. Take advantage of this by bridging from their uses of shortcuts to the later development of formulas.

There are many complex processes going on in teaching measurement to which teachers must attend. Understanding how students learn and perceive measurement is the first step.

Reference

Wilson, P., and R. Rowland. "Teaching Measurement." In *Research Ideas for the Classroom: Early Childhood Mathematics*. Vol. 1. Reston, Virginia: National Council of Teachers of Mathematics and MacMillan, 1993.

What Research Says About What Children Know About Measurement

Think about all the measurements you consider in your life

from the time you awake,

to the time you arrive at school.

*Different measures often have different **standard** units.*

*Sometimes we use **nonstandard** measures, such as cans, desks, and paper.*

*Sometimes we **estimate** measures . . . sometimes we make **precise** measures.*

Sometimes we use **comparisons:** "It is cooler today than yesterday."

Sometimes we use **averages:** "I drive an average of 30 miles a day."

Some measures are **rates:** kilometers per hour, miles per gallon.

Children develop spatial concepts at two levels.

*They develop **perception** of spatial concepts from physical contact with objects—playing with blocks, tracing shapes, doing puzzles, walking along a wall.*

*They develop **representations** of spatial concepts by using memory and imagination.*

As children perceive and represent shapes, their perceptions change, improving their representational skills.

Being actively involved in measuring helps students to construct ideas.

The goal is to help students eventually construct abstract ideas such as unit, perimeter, area, and rate.

After construction, students need to reflect on their ideas by talking and writing about them.

We have been taught and tested about measurement using a collection of properties, such as:

$$1 \text{ yard} = 3 \text{ feet}$$
$$\text{perimeter} = 2(L + W)$$

*Today's students must **construct ideas** for themselves instead of memorizing rules.*

They can then discuss their ideas and move toward applying the new ideas to real-world problems.

Measurement has three fundamental components:

Conservation

Transitivity

Unit

Conservation

An object maintains the same shape and size if it is moved or subdivided into parts.

Test for conservation of length
Materials: 2 straws

Test for conservation of area
Materials: 2 different-colored pieces of construction paper, scissors

Test for conservation of volume
Materials: 3 plastic containers (2 round, 1 rectangular), water

Transitivity

If A is less than B
and B is less than C,
then A must be less than C.

If $A < B$
and $B < C$,
then $A < C$.

Unit

Students need to understand what attribute is being measured.

Length: *A unit that is 1-dimensional*

Area: *A unit that is 2-dimensional*

Volume: *A unit that is 3-dimensional*

Students also need to understand how the chosen unit influences the number of units.

Students cannot focus on several dimensions, the number of units, and the size of the units all at the same time.

They need your help!

*Some students attach too strong a meaning to **number** at the expense of **measurement**.*

The process of learning to measure can be divided into five sequential steps designed to span grades K–5.

1. *Identify the property to be measured.*

2. *Make comparisons.*

3. *Establish an appropriate unit and process for measuring.*

4. *Move to a standard unit of measurement.*

5. *Create formulas to help count units.*

INVESTIGATION 11

Accuracy in Measurement

Overview

Even the most carefully taken measurements involve some uncertainty. The average of several measurements is less variable (more reliable) than a single measurement, which is why it is important to take repeated measures. A measurement is considered to be reliable if, using the same units, it can be consistently replicated. Bias, as well as inaccurate measurement techniques, affect reliability adversely. *Bias* refers to systematic deviation of the measured result from the "true value" that perfect measurement would produce. Inaccuracy and bias are both sources of error in statistical investigations and can jeopardize results. They must be reported by researchers in experimental studies.

Assumptions

Participants are able to recognize that variations in measures occur.

Goals

Participants explore the concept of measurement. In particular, they

- recognize that errors might be introduced in the course of the data-collection process

- recognize the implications of data-collection errors in the analysis of data

Reference

Moore, D. S. *Statistics: Concepts and Controversies.* 3rd ed. New York, New York: W. H. Freeman and Co., 1991.

Materials

Inchsticks, calculators

Handout 11.1

Developing the Activity

Make connections between participants' previous work and sources of error in the data-collection process.

In several investigations, we have noted that variations in measuring and measurements occur. For example, in measuring foot lengths, variations were a result of the way we interpolated, or estimated, fractions of a unit. We also had variations due to differences in measuring procedures, such as whether we measured the length of the foot with shoes on or shoes off.

Variations in measurement occur for many reasons. Sometimes the way one person reads a ruler differs from the way another reads it. If you round a measurement down and I round it up, our data will have variations that may need to be controlled or prevented.

The data-collection process involves many opportunities for error. In scientific data analysis, we want to minimize sources of error as much as possible. One way to minimize errors in the collection of data is to collect multiple or repeated measures of the attribute.

Pose the Question

What types of problems arise when collecting data? What does it mean when we say that the measures taken in collecting data are unbiased *and* reliable?

Collect the Data

Discuss issues related to accuracy in measurement. Accuracy in measurement involves two components: bias and reliability.

- A measurement is *unbiased* to the extent that it does not overstate or understate the actual value of the attribute being measured.

- A measurement is *reliable* to the extent that repeated measures of the same values of the attribute yield the same (or approximately the same) results.

- The results of repeated measurements of the same values will reflect a distribution of measures. The average of these measures will be more reliable than a single measurement.

- Bias and reliability apply to all measures. Examples include mass, length, SAT scores, classification of employment status, and enrollment numbers for first-time college students.

Divide participants into teams of four. Have teams work together to complete Handout 11.1.

Analyze the Data

Have teams discuss the questions on the handout.

Interpret the Results/Summary

Ask participants for their reactions to this activity and to share any experiences they have had with students related to issues of accuracy in measurement.

HANDOUT 11.1

Accuracy in Measurement Problem Sheet

1. Use your inchstick to measure the line below to the nearest hundredth of an inch (such as 5.25 inches or 5.41 inches). In making this measurement, you need to estimate what portion of the line extends into the distance between the 5- and 6-inch marks. Careful measurements usually involve uncertainty. The uncertainty here is magnified by the use of an instrument that is divided only into inches.

   ▬▬▬▬▬▬▬▬▬▬▬▬▬▬▬▬▬▬▬▬

   a. What is your measurement?

   b. Record the measurement each of your team members found.

   c. What can you say about the measurements your team found? Is measuring with an inchstick reliable (that is, not variable or imprecise)?

   d. Calculate the mean length found by your team.

   e. What would you expect to find if you asked everyone in the Teach-Stat workshop for their measurement of the line? Why?

f. When measuring a line segment, students often place the *true edge* of their ruler—instead of the 0 mark—on the left end of the line segment. This practice causes *bias*. Suppose you measured the line segment with a standard ruler by placing the left end on the 0 mark and *every other person in your team* measured the line segment by placing the *true edge* of the ruler at the end of the line segment. Would your measurements be biased? Why? Would your measurements be reliable? Why?

2. Suppose each participant was asked to measure her or his resting pulse rate. They use one of two methods.

   Method 1: count the number of heart beats for 10 seconds and multiply by 6 to get beats per minute

   Method 2: count the number of heart beats for 30 seconds and multiply by 2 to get beats per minute

   a. Is either method more reliable? Why?

   b. Is either method more biased than the other? Why?

From: *STATISTICS: CONCEPTS AND CONTROVERSIES* by Moore. Copyright © 1991 by W.H. Freeman and Company. Used with permission.

INVESTIGATION 12
How Close Can You Get to a Pigeon?

Overview

Participants have an opportunity to integrate measurement concepts and to apply the process of statistical investigation to solve a problem. The activity also offers them with a chance to get outside the classroom and to get acquainted with each other.

Assumptions

Participants are familiar with the four parts of the statistical investigation process and are aware that in collecting measurement data, a plan must first be specified.

Goals

Participants explore the process of statistical investigation. In particular, they

- define the question

- establish a method and formulate a plan for measurement and data collection

- collect data

- organize and analyze data

- draw conclusions

- report methodology and findings

- build theories based on their data

Reference

Corwin, R. B., and S. J. Russell. *Measuring: From Paces to Feet*. Palo Alto, California: Dale Seymour Publications, 1990.

Materials

Metric and inch-based measuring tapes including one 10–20 meter retractable tape per team, chart paper, colored markers, colored stick-on dots, calculators

Fun Facts

Birds tend to stand on one foot while sleeping to give the other a rest. This behavior also helps prevent body-heat loss from a leg not covered (insulated) with feathers.

To preserve more body heat, birds also sleep with their heads tucked under their wings or feathers.

Developing the Activity

Participants first consider the question, then collect and analyze data.

Pose the Question

What is meant by "comfort distance"?

Comfort distance is the point at which an animal becomes uncomfortable with the close physical presence of another animal.

Have you ever had an experience where you felt there was not enough "space" between you and another human being?

Did you know that personal space measurements are often a factor in restaurant design?

How might we conduct an investigation to determine human comfort distance or space? What risks might be involved in such an investigation?

Although we are not social scientists, we will be conducting an investigation into comfort zones, but not with humans. The subject of our investigation will be pigeons! The question we will investigate is this: How close can you get to a pigeon?

How might we collect data?

What problems might we encounter (for instance, having no pigeons nearby to study)?

What tools will we need?

What about using lures, such as bread? Would the use of lures change the question from "How close can you get to a pigeon?" to "How close can you get a pigeon to come to you?"

Where can we collect our data?

What constitutes an adequate sample?

As a group, come to a consensus on methodological issues.

Collect and Analyze the Data

Divide participants into teams of three or four. Give directions to nearby spots that are known to be places where pigeons flock. Participants should have enough time to collect, represent, and analyze their data.

Interpret the Results

Have teams share stories about the collection of the pigeon data.

Record and display data. Notice whether or not measurement units are standard.

Once each team has presented their data, have each participant work with a partner to synthesize the data and come up with responses to the following questions:

- *Are there any patterns in the data?*

- *What do the data show?*

- *Are the results clustered?*

- *What would you expect if you went outside right now and observed a pigeon? How close could you get?*

- *Can you think of other questions for investigation that are outcomes of this study?*

- *How does this investigation illustrate some of the key data-analysis concepts we have presented thus far in the workshop?*

INVESTIGATION 13

Family Size

Overview

The question in this investigation is deceptively simple: How big are our families? The primary difficulty in answering the question is defining *family*. Participants are expected to raise questions concerning types of families that may not be represented in the group, comparisons of current families with those of parents or grandparents, and so on. The problem is posed to the group as a whole, after which the use of smaller groups of people is an option for purposes of discussion. Facilitating the discussion in a way that encourages interaction among participants is critical to the success of this activity.

The formulation of clear questions is crucial in any statistics activity. Participants will have many opportunities to apply their question-formulation skills in the remainder of the workshop. Formulating clear questions may be even more critical in ill-defined areas. Thus, there may be an important transfer of these skills to interdisciplinary content such as social studies and literary analysis.

Circumstances within the group may occasionally make this activity inappropriate. For example, a participant may recently have experienced a death in the family and find discussions of family very painful. In such a situation, substitute a different question, such as, How many doors do we have in our living areas? (See "Extension" at the end of this activity.)

Assumptions

Teachers will be familiar with a variety of personal relationships (parents, siblings, aunts, cousins, and so on) and with some types of representations (for example, bar graphs and tallies). Some instruction may be needed with respect to some types of representations (line plots, stem-and-leaf plots, and so on) and to comparing types of representations (for example, What is an example of types of information that can be read from a stem-and-leaf plot as opposed to a bar graph?).

Materials

Pens, paper, graph paper, colored stick-on dots, interlocking cubes, colored markers, calculators

Transparencies 13.a through 13.c

Handout 13.1

Family Size 107

Goals

Participants explore the process of defining the question. In particular, they

- learn how posing a focused question is critical to the process of statistical reasoning

- learn how questions may be refined more than once during an investigation

- understand various ways of representing data

- understand the appropriateness of various representations

- describe the shape of data

- identify different interpretations of data

References

Russell, S. J., and R. B. Corwin. *Statistics: The Shape of the Data*. Palo Alto, California: Dale Seymour Publications, 1989.

Professional Standards for Teaching Mathematics. Reston, Virginia: National Council of Teachers of Mathematics, 1991.

Developing the Activity

Participants will consider how to carefully define terms when formulating questions.

Pose the Question

Some people are very curious about the make up of families. Who might want to know about family size? Do we all come from the same size families? How big are our families today? The task for you today is to investigate the sizes of our families.

Suggest that participants try to use their "third eye"—be participant observers—to observe what the discussion leader does during the ensuing exploration of this problem.

Try to reflect on what is happening during the activity. Later, we can discuss what the investigation leader thought were important points of discussion.

Teacher Notes

Unlike many children, most adults have at least two families to consider: their family of origin and their current "family."

As the investigation leader, you may want to restate the general question—How big are our families?—and then ask:

What data must we gather to answer our question?

Participants may initially view this as a simple question; however, as several suggestions are made about what data to gather, it will become clear that there may be lots of ambiguity in what participants want to count as *family*.

If participants don't do so spontaneously, pose this question:

What people might be included in a family?

Brainstorm together for a few minutes, recording suggestions on the overhead projector or board.

How will we decide which people to include? Should we ask smaller groups to discuss options, or should we decide as a whole group?

Participants may want to let subgroups work independently for a few minutes and then report back to the whole group; however, the group as a whole must reach consensus during this initial discussion. The discussion may take 15 to 30 minutes. Let it continue as long as it seems profitable, trying to avoid having it drag on too long while permitting participants to deal with the ambiguity of having no "one, right" answer. You may find it easier to facilitate the discussion by asking such questions as:

If we used this definition of family (indicate the current definition under consideration), *who would be counted? Who would be left out? Is using this definition of family a problem for anyone?*

Based on what we are thinking now as a group, who would you count in your family? Is there anyone who isn't clear about who to count now?

Do we want to count families in more than one way and see what differences there are in the data?

Before having participants collect data, you may want to have them reflect on the nature of the discourse during the question-formulation process. Possible discussion topics to consider include

- Why did I let you struggle for so long?

- Why didn't I simply give you a definition of family?

- Why didn't I suggest that we take a vote?

Family Size

- How would you support such a discussion with your students?

- What does this activity so far have to do with statistics?

Collect the Data

Let the group gather the data it needs in the way participants think will be most efficient.

Analyze the Data

Divide the group into teams of three to six people, and ask each team to develop several representations of the data. Have a variety of materials available including graph paper, plain paper, and interlocking cubes.

Deciding which representations to use may not be easy for participants and will probably elicit considerable discussion. Monitor each team, and keep all participants engaged in the process of making representations. It is not unusual for participants to take turns making representations. If you find this happening, encourage participants who are watching to work on their own representations.

Ask teams to discuss among themselves the appropriateness of each type of representation.

What are the advantages to using each representation you have developed? What are the disadvantages?

Ask each team to present its representations. Participants will probably have created a variety of representations; for example, real graphs, towers of interlocking cubes, pictographs, bar graphs, line graphs, line plots, and computed values (that is, mean, median). Direct discussion so that a wide variety is presented. Focus attention on comparison of the representations.

What are the common features of these representations? What are the unique features of these representations?

Interpret the Results

What can we now say about our family sizes? What is typical for us?

What new questions do these results suggest? What would you like to know next about family size?

Brainstorm for a few minutes. Participants may focus on characteristics of subgroups of people (for example, gender, geography), on generational differences (such as participants versus their parents or grandparents), or on differences between participants and other groups not represented in the workshop (for example, single-parent families). Encourage participants to move in whatever direction seems most interesting.

Work for consensus on one or two questions that could be investigated as follow-up questions.

Summary

If there is time, ask participants to select one of the new questions and try to answer it, following the same process used earlier. They should be sure all terms are clearly defined, gather the data, represent the data in a variety of ways, and interpret the results.

It is important to point out that this version of the *Family Size* activity is more involved, and probably richer in ideas, than the version that might be presented to elementary students. This activity was developed from the perspective of adults, though the topic of family size can be dealt with profitably by children in grade 3 or higher.

Teachers who have used this activity with elementary-grades students have not experienced significant difficulty in having them discuss a wide variety of kinds of families. Children generally seem sensitive and responsive to differences among the lifestyles of their peers. Teachers should expect that children will not raise nearly as many issues as adults in the process of reaching consensus on the definition of *family*.

Facilitate a discussion about the relationship of this activity to the material addressed in *Professional Standards for Teaching Mathematics* (NCTM 1991).

- *How does this activity encourage the development of discourse, both among students and between teacher and students?*

- *Is the problem a worthwhile mathematical task?*

Encourage discussion about how changes in the problem setting might enhance the development of any of the Professional Standards. You may want to ask specific questions about teaching.

- *Which components of the process do you think are easier to facilitate? More difficult to facilitate?*

Teacher Notes

This problem will generate a great deal of discourse. Many participants will want to express their views during the process of defining family. Be alert for the roles different people may play; for example, which people are more dominant in the discussion?

Try to monitor your own discourse, particularly regarding the amount of direction you give to the discussions. How long are you comfortable letting participants flounder? When do you interject with a focusing comment or question?

If there is time, let participants discuss these important issues.

Bar Charts and Bar Graphs

You may want to address bar charts and bar graphs as ways of representing information. Refer to Transparencies 13.a, 13.b, and 13.c:

- *Family Size Bar Chart: Ungrouped Data (Unordered)*
- *Family Size Bar Chart: Ungrouped Data (Ordered)*
- *Family Size Bar Graph: Grouped Data*

A bar *chart* shows a bar for each data item; a bar *graph* provides a frequency distribution of the data. It is important for teachers to focus on the differences between the representations, specifically looking at the information provided by each of the axes and what information the size of the bars provides.

On a bar chart, each bar shows the actual data for each member of the data set; for example, a bar of height 3 (along the y-axis) means there are three children in the family of a named person (along the x-axis). Taller bars represent larger numbers of children.

On a bar graph, each bar shows the number of members of a data set having a given value; for example, a bar of height 3 (y-axis) means there are three families with a given number of children (x-axis). Each unit in a bar is the same size, so the bars represent counts and not family size or number of children.

We want to use bar graphs for statistical purposes and should avoid use of the simpler bar chart if we are to consider such concepts as shape and grouping of data.

Extension

As an extension of this activity, ask participants how many doors are in their homes, apartments, or other living areas. Have them visualize walking through their homes and counting doors. Record guesses on the overhead, and then discuss what was counted.

Someone will probably mention cupboard doors, oven doors, refrigerator doors, and the like. Again, you will need to reach consensus on what will count as a door. (If the group suggests something like "flat surface connected to a frame by hinges" as a definition for *door,* ask if a toilet seat counts as a door!)

Request that participants count doors for homework. Display and analyze results, drawing some conclusions about the number of doors participants have in their homes.

Family Size Bar Chart: Ungrouped Data (Unordered)

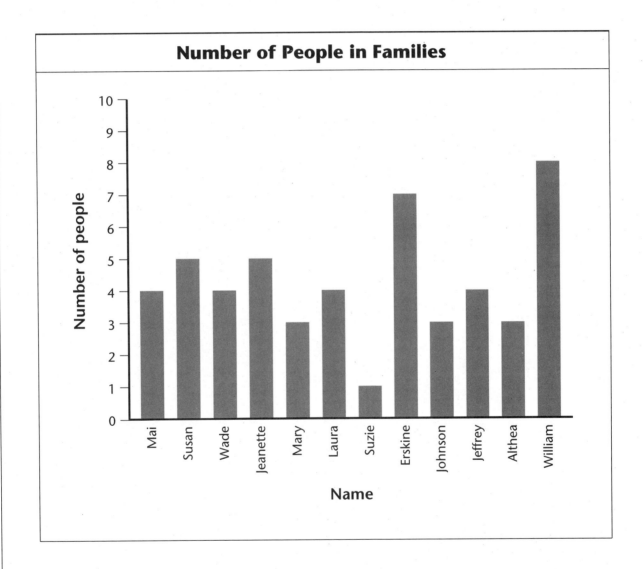

Family Size Bar Chart: Ungrouped Data (Ordered)

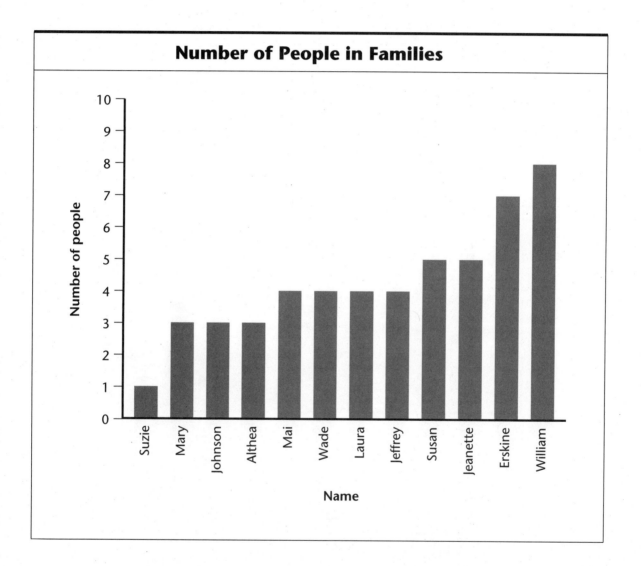

Family Size Bar Graph: Grouped Data

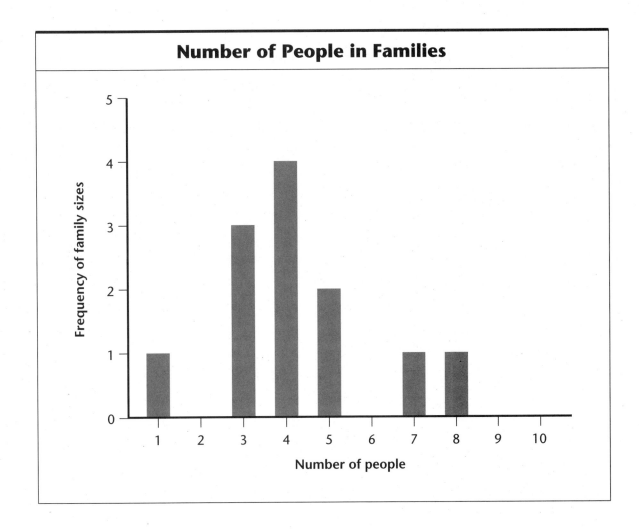

HANDOUT 13.1

Family Size

1. How big are our families? Who might want to know?

2. What data do you need to answer the question?

3. How might you define *family*? Which definition is best? Why?

4. Gather the data you need to answer the question for this group. Record the data below.

5. Make several different representations of the data.

6. What new questions come to mind? What would you like to know next about family size?

7. How does this activity encourage discourse, both among students and between teacher and students?

8. Is the problem context for this activity a worthwhile task?

INVESTIGATION 14
Median: More Than Just the Middle of the Data

Overview

To focus on the use of the median as a summary statistic that can be used to describe data, participants explore four problem situations. The four problems are

- What is the median pencil length in a handful of pencils?

- What is the median number of people in our group's families?

- What is the median number of raisins in half-ounce boxes?

- Under how many different roofs have we lived?

The first problem provides an informal means for quickly generating a data set and then, by literally "working toward the middle," participants locate the median. The second problem revisits earlier data participants have collected; they physically represent the data using interlocking cubes, order the data, and locate the median, again by working toward the middle. The third problem revisits another set of data participants collected earlier; they order the data and locate the median. The final problem engages them in the full statistical-investigation process, with the goal being to compare their group's data with data provided by two other groups. Median and range are useful tools for making quick comparisons and often provoke discussion about ways to compare different sets of data.

Assumptions

Participants have had some introductory work with data analysis. In particular, they are familiar with the use of line plots to represent data, are comfortable with the concept of the shape of the data, and have explored what's typical about a set of data.

Materials

Interlocking cubes, grid paper, scissors, calculators

Transparencies 14.a, 14.b

Handouts 14.1, 14.2

Goals

Participants will understand the concept of median. In particular, they will

- depict the identification of the median using objects, employing the "working toward the middle" strategy to locate the median.

- pictorially depict the identification of the median using numbers, ordering the numbers, and identifying the median by using these strategies:

    a. working in toward the middle

    b. folding paper strips

    c. determining how many data values represent half of the total number of data values

- work backward from a line plot to disaggregate data and locate the median

References

Russell, S. J., and R. B. Corwin. *Statistics: The Shape of the Data.* Palo Alto, California: Dale Seymour Publications, 1989.

Fey, J., W. Fitzgerald, S. Friel, G. Lappan, and E. Phillips. *Data About Us.* Palo Alto, California: Dale Seymour Publications. Forthcoming.

Developing the Activity: Part 1

Participants will order sets of data and explore the meaning of *median*.

Pose the Question

Have groups of participants pool their pencils, pens, crayons, and so on, gathering at least 15 writing tools. If groups have fewer than 15 items, team groups together so that each group has at least 15.

How many writing tools does each group have?

You may want to record this information somewhere for future reference. Make sure some groups have an odd number of items and some have an even number of items.

Have groups order their collections from shortest to longest.

Collect the Data

Working in from the ends of your collection, pair the shortest and the longest writing tools and set them aside. Continue this until only one or two items remain.

Analyze the Data and Interpret the Results

After this "working toward the middle" strategy, some groups have just one item left. What can you say about the number of items that come before this writing tool when the set is ordered from smallest to largest? The number of items that come after this writing tool?

The number of writing tools before and after this middle item are the same. The length of the writing tool that remains marks the halfway spot, or the middle location, of the lengths of the items in the ordered set.

Suppose I had 39 pencils, and this pencil [hold up a pencil] *was the middle length. What do you know about how many of the other pencils would come before or after this pencil in an ordered arrangement?*

In an ordered arrangement, 19 pencils would come before the middle pencil, and 19 would come after the middle pencil.

What can you tell me about the lengths of the 19 pencils that come before this pencil and the 19 that come after this pencil?

The 19 pencils that come before the middle pencil must be shorter or equal in length to it; the 19 pencils that come after must be longer or equal in length to the middle pencil.

For those groups with one writing tool remaining, we say that the length of that writing tool is the median *length in the ordered set of data. It separates the ordered set of data exactly in half.*

What about the groups that have two items left? How can we identify a median length for a writing tool that separates the data set in half?

You want participants to discuss the need to find the length that would be halfway between the lengths of these two items—the *average* of the two lengths. They will need to measure the two writing tools and compute the average.

For the data sets with two writing tools remaining, is the median length always the length of one of the writing tools in the data set? When will this be true? When will this not be true?

This will be true only when the lengths of the two items are the same. If the lengths are not the same, the median length will be an average of the two lengths (and thus will not be the length of one of the actual writing tools).

Suppose I had 40 pencils and the median length of the pencils was 15 cm. What do you know about how many of the remaining pencils come before or after the median in an ordered arrangement?

The lengths of 20 writing tools are before the median, and the lengths of 20 writing tools are after the median.

What are some possible lengths of the twentieth and twenty-first pencils in the ordered arrangement?

The pencils will either both be 15 cm long, or they will have two different lengths with a mean of 15 cm (for example, 14 cm and 16 cm, or 13.5 cm and 16.5 cm).

What are possible lengths of the lower 20 pencils?

Each of the lower 20 lengths may be less than or equal to 15 cm.

What are possible lengths of the upper 20 pencils?

Each of the upper 20 lengths may be greater than or equal to 15 cm.

Remember that the median *is the number that divides an ordered data set in half. It may also be the middle number of a data set, but it does not have to be.*

Developing the Activity: Part 2

Participants will use a visual model to find the median of a data set.

Pose the Question

Returning to our data on family size, how can we find the median family size for our group?

Collect the Data

Have participants re-create the data about the number of people in their families by making cube towers. They can then place the towers in an ordered display for everyone to see.

Analyze the Data and Interpret the Results

Participants can apply a "working toward the middle" strategy to determine the median family size. Again, you will want to consider a situation in which there is an odd number of families and one in which there is an even number of families (perhaps with and without your data).

Suppose a class had 24 children and the median family size was 4 people. What do you know about the how many of the remaining students' families come before or after the median in an ordered arrangement of the data?

You know 12 families precede and 12 families follow the median. The median is the mean of the twelfth and thirteenth family sizes in the ordered set of data.

Describe what the data might look like if it were represented with cube towers. What might you expect to see displayed? Why?

There would be at least two family sizes of four people or whose mean family size was 4 people. These would be the twelfth and thirteenth data values in an ordered arrangement of the data. The remaining family sizes before the median will be less than or equal to 4 and those after the median will be greater than or equal to 4.

Developing the Activity: Part 3

Participants will generalize a strategy for finding the median for an ordered set of data.

Pose the Question

Returning to our data about the number of raisins in a half-ounce box, how can we find the median number of raisins in a box for our group?

Introduce this task using the examples shown on the next page.

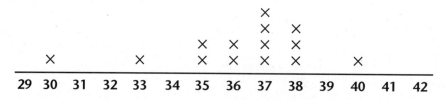

Provide each person with graph paper that has enough squares so the number of raisins in each box can be listed in order from smallest to largest:

| 30 | 33 | 35 | 35 | 36 | 36 | 37 | 37 | 37 | 37 | 38 | 38 | 38 | 40 |

Participants can cut out the strip of values and fold it in half. In this example, there are an even number of data items (boxes of raisins), so the fold falls on the line between two boxes.

fold line

| 30 | 33 | 35 | 35 | 36 | 36 | 37 | 37 | 37 | 37 | 38 | 38 | 38 | 40 |

How can we determine the median number of raisins using this strategy?

Suppose we opened three more boxes of raisins and counted these numbers of raisins: 35, 35, and 35.

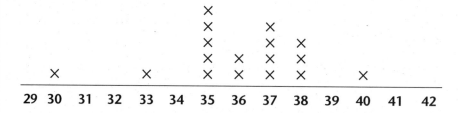

Make a new strip of data, and find the new median using the "fold in half" strategy. What do you notice this time?

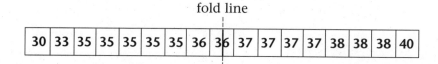

Collect the Data

Have participants re-create the data about the number of raisins in the boxes they opened.

Analyze the Data

Have participants locate the median for their raisin data set. Encourage them to find different ways to do this, using both strategies they have already seen and invented other strategies.

Interpret the Results

Have participants share their strategies.

Participants may discuss strategies of determining what is half of the total number of data values. This number permits them to quickly identify which data value or pair of data values in an ordered set of data may be used to determine the median.

This next question will help develop this strategy.

Suppose I had opened 43 boxes of raisins. What information would you need to find the median of the data?

Initially, participants may indicate that they would need all the data. Hopefully, as the discussion progresses, they will realize that if the data are in order from smallest to largest, they need only the twenty-second data value, as it is the median value. Encourage them to formalize this strategy, describing the need to find "half the data values plus one more" to be able to identify the median of an *odd* number of data values *if* the data are ordered.

Suppose we had opened an even number of boxes of raisins. What information would you need to find the median of the data?

You may want to offer specific examples (for example, 24 boxes or 36 boxes). Encourage participants to generalize a strategy, describing the need to find "half the number of data values" and locate the data value at this point.

Developing the Activity: Part 4

Participants will engage in the entire statistical-investigation process, comparing data they generate to existing sets of data.

Pose the Question

In this activity, participants consider the following questions:

How many different roofs have you lived under since you were born?

Teacher Notes

Here is one strategy to determine the median.

1. All data values should be placed in order from smallest to largest.

2. Count the total number of data values.

3. If the number of data values is *odd*, the median is the middle data value.

4. If the number of data values is *even*, find the two data values in the middle of the list. Compute the average of these two values to determine the median.

How does our data compare with data from two other groups of people?

Participants will have to discuss what is meant by "lived under." One group decided that you had to have received mail at that location. Participants also need to clarify any questions that arise. For example, would a roof count more than once if they had lived under it at different times in their lives?

Collect the Data

Have participants display their data using a line plot.

Provide copies of Handouts 14.1 and 14.2, *Roofs 1* and *Roofs 2*, for each participant.

Analyze the Data

How can we use what we know about the shape of the data to help us compare the three sets of data?

This question will encourage participants to look at where the data cluster, what kinds of gaps there are, and the spread of the data.

Does finding the median and the range of each of the sets of data provide information that will help us compare the three sets?

Interpret the Results

In comparing the three sets of data, what kinds of summary statements might we make?

What questions do we have about the data sets that may affect how we choose to interpret the data?

Summary

Participants have been introduced to the concept of median and have been exposed to various strategies for finding the median.

Teacher Notes

The median is a summary statistic that can be used to describe data. It is the value that separates a set of ordered data in half, so that 50 percent of the data are in the lower—or upper—halves of the data set.

The median may not be distinctive; there may be more than one data item with the same value as the median. If so, the ordering of the data arbitrarily determines which of these equal values are considered to be in the lower—or upper—halves of the data set. Also, the median may not be an actual value in the data set but instead be computed by finding the average of the two middle values in an ordered set of an even number of data items.

Roofs 1

How Many Different Roofs Have You Lived Under?

```
                    X
        X           X
        X       X   X
  X     X       X   X       X X
  X X   X X X   X X X       X X X
X X X X X X X X X X X X     X X X X
1 2 3 4 5 6 7 8 9 10 11 12 13 14 15 16 17 18 19 20 21 22
```

Number of roofs

Roofs 2

How Many Different Roofs Have You Lived Under?

```
                                                          x
                              x
                        x  x  x
                        x  x  x
                     x  x  x  x
                     x  x  x  x
                  x  x  x  x              x
         x  x  x  x  x  x  x     x                    x
   x  x  x  x  x  x  x  x  x     x        x     x     x         x
   ─────────────────────────────────────────────────────────//──────
   1  2  3  4  5  6  7  8  9 10 11 12 13 14 15 16 17 18 19 20 21 22  43
                           Number of roofs
```

Roofs 1

How Many Different Roofs Have You Lived Under?

	1	2	3	4	5	6	7	8	9	10	11	12	13	14	15	16	17	18	19	20	21	22
		X	X	X		X	X					X	X					X	X			
			X	X	X	X	X				X	X	X					X	X	X	X	
						X	X					X	X						X	X		

Number of roofs

HANDOUT 14.1

HANDOUT 14.2

Roofs 2

How Many Different Roofs Have You Lived Under?

```
                              x
                        x     x
                  x x x x     x
                  x x x x     x
                x x x x x     x
              x x x x x x     x
            x x x x x x x     x     x
        x x x x x x x x x     x   x x
------------------------------------------
1 2 3 4 5 6 7 8 9 10 11 12 13 14 15 16 17 18 19 20 21 22 // 43
```

Number of roofs

INVESTIGATION 15

Types of Data: A Minilecture

This minilecture is on the various types of data and scales and the statistics that can be used with them.

Transparency 15.a

Measurement is merely a means of assigning values to a collection of data. *Analysis* is the process of reflecting on those values to obtain new information about the data and, possibly, to gain a better understanding of the population from which the data were collected.

The theory of measurement consists of a set of separate theories, each concerning a distinct *level of measurement*. These levels of measurement are often referred to as *types of data*.

A basic understanding of the level of measurement—the type of data—used is important, because the level of measurement affects how the data can be analyzed. Assuming a quality that is inappropriate for a set of data can lead to analysis results that are incorrect or even meaningless. It will be easier to understand this concept after looking at a few examples.

There are several ways to classify types of data. We will consider four categories: nominal, ordinal, interval, and ratio.

Transparency 15.b

Data classified as *nominal* demonstrate the weakest scale or level of measurement. The values assigned to nominal data may or may not be numbers, and they serve only to identify the data.

This is often referred to as a *classification* or *categorical* measure. Two identical values indicate that the same value occurred twice or that two values occurred that are indistinguishable. Two different values indicate two different pieces of data that can be distinguished from one another.

Materials

Transparencies 15.a through 15.f

Examples. Colors, license plate numbers, football jersey numbers, nationalities, gender, breed of cow, political affiliation, states of the United States, brands of 12-ounce cans of whole-kernel corn

Properties. Values in a nominal scale of measurement can only be partitioned or separated into mutually exclusive collections. One may also think of saying the data may be "sorted" provided that *sorted* means split or separated rather than ordered—for example, from bad to good or small to large. The only possible relationship for comparing data values in the nominal scale of measurement is equality (=) or inequality (≠).

Transparency 15.c

An *ordinal* scale has all of the features of the nominal scale, but additionally, the objects in one category are not only different from the objects in another category but also exhibit a more specific relationship: they are bigger, colder, older, more appealing, and so on. The measurements have an *order* to them; hence the term *ordinal* data. The relators > and < can be used to compare ordinal data.

Examples. Ranks in employment: sergeant, corporal, private; style points assigned in gymnastic events; attitude scales: strongly agree, agree, undecided, disagree, strongly disagree; intensity: low, medium, high; numerical ranks: first, second, third, . . . , last

Properties. Values in the ordinal scale have all of the properties of values in the nominal scale, with the additional property that the values can be ordered.

Transparency 15.d

An *interval* scale has all of the features of the ordinal scale, but additionally, distances or differences between values on the scale are constant. This is the first scale to include the concept of a unit of measurement. If two values differ by a unit at one end of the scale, two other values that differ by a unit elsewhere in the scale exhibit the identical difference.

Examples. Fahrenheit degrees, Celsius degrees, weight (measured on a spring scale), time

Properties. This is a truly quantitative scale. Differences in values in the interval scale have many of the properties of arithmetic. Mean, standard deviation, and correlation are applicable in the analysis of these data. None of these measures have meaning for nominal or ordinal scales. A common analysis error is to average ordinal data. Without a defined unit of measurement, an average is meaningless.

Transparency 15.e

A *ratio* scale has all of the features of the interval scale, but with the additional property of having a true zero point at the origin. The ratio of any two values in this scale (as in A/B) is independent of the unit used in the measurement.

Examples. Count, length, area, volume, Kelvin temperature, mass (measured on a balance), duration

Properties. The existence of a true zero point along with the features of the interval scale allow us make such statements as "20 is twice as much as 10." For our purposes, you may assume that any statistical procedure can be used on values in this scale.

Transparency 15.f

When we use statistics or representations of data, we must take the type of data we are analyzing into account.

Mode, Median, and Mean

Consider the evaluation of the mode, median, and mean for a set of data.

If the data are measured on a nominal scale, the mode is an appropriate statistic; it merely counts the number of observations in each category. The median cannot be used because there is no one way to order the data. The mean is similarly inappropriate.

Data measured on an ordinal scale can be described by a mode if repeated measures are possible. The median is a very useful statistic to describe ordinal data. The mean should not be considered in ordinal data because it assumes a measurement scale that has a defined, constant unit of measure.

The mode, median, and mean can all be used to describe data measured on either the interval or ratio scales. Often there are no (or few) repeated data values (or ties) with these scales, since an infinite number of different values are possible. When there are few repeated values, the mode loses its meaning.

Discrete Data and Continuous Data

Another aspect of types of data that is of interest is the idea of discrete as opposed to continuous data. *Discrete* data implies that only certain values are possible. If 1 and 2 are possible, no values between 1 and 2 can occur. Data on a *continuous* scale can have any value between any other two values. If 1 and 2 are possible, then 1.5, 1.1, and 1.76342 are all possible even if they do not occur in the data set.

A continuous scale of measurement has an infinite number of possible values. Surprisingly, the discrete scale could be infinite, yet it is often finite.

Nominal data is frequently measured on a finite, discrete scale. Ordinal data could come from either a discrete or continuous scale. Interval and ratio data are frequently measured on a continuous scale. There can be exceptions to these guidelines.

Multivariate Data

Multivariate data can add interesting aspects to the concept of types of data. Consider the example of measuring the height of students in a middle-grades class. The actual heights are measured on a ratio scale. Assume that, at this age, it is possible that one gender has a different stature than the other. Thus the gender of each student is also collected and paired with the height. Gender is naturally measured on a nominal scale.

Types of Data

What are they?

Why are they important?

Measurement: *Assigning values to observations*

Analysis: *Reflecting upon these values to gain information*

Theory of measurement: *The level of measurement or type of data*

Nominal Scale

Weakest level of measurement

Categorical measure

Examples: *fur, color, gender, nationality, religion, shape of leaves*

Property: *equal or not equal*

Ordinal Scale

Stronger level of measurement

Examples: *ranks, attitude survey, Olympic diving scores, finishing place in a race*

Property: *less than or greater than*

Interval Scale

Stronger level of measurement

Examples: Fahrenheit degrees, Celsius degrees, weight, time

Property: some unit of measurement exists

Ratio Scale

Strongest scale of measure

Same as interval scale with the addition of a true zero point

Examples: count, length, area, volume, Kelvin degrees, mass, duration

Property: 8 is twice 4, etc.

Scale and Measures of Center

Mode: The most frequently occurring value in a data set

Median: The middle value in an ordered set of data

Mean: The value if each data item in the data set were the same value

Use of Statistical Measures

	Mode	Median	Mean
Nominal scale	×		
Ordinal scale	×	×	
Interval scale	×	×	×
Ratio scale	×	×	×

INVESTIGATION 16

About Us Revisited

Overview

Participants return to the work they completed in Investigation 1, *About Us,* and look at both the questions they asked and the representations they made to display their data.

Assumptions

Participants have completed the *About Us* investigation as an introductory activity. The materials they produced during that investigation are still on display or available to be viewed.

Goals

Participants reflect on the types of data they collected as part of the *About Us* investigation. In particular, they

- identify the types of data that were collected

- determine which measure—mode or median—is applicable for describing the center of each set of data

- discuss what it means to talk about the spread of the data as it relates to different types of data

Developing the Activity

Your work with participants will depend on the results from the *About Us* investigation. Use participants' data and graphs to consider the kinds of data and related statistics as is demonstrated in the three examples that follow—explorations that were conducted in Teach-Stat workshops.

Categories of Alternative Professions (categorical data)

The team doing this exploration asked each participant, What profession would you rather be in? They collected a list of professions and, to group them, coded the answers as (H)uman Resources, (M)ath, (S)cience, (A)rts, (Sp)orts, (O)ther, and (U)ndecided.

Materials

Graphs and other materials participants created during the *About Us* investigation, calculators

Handout 16.1

List of Professions

H	teacher (2)	S	pharmacist
H	social worker	O	homemaker (3)
Sp	tennis pro	S	biologist
O	business executive	S	obstetrician
A	interior decorator	A	comedian
A	writer	M	banker
O	shrimper	A	florist
U	don't know	H	newscaster
O	bird watcher	M	retail sales
O	Navy	O	work for a publisher
M	accountant (2)	O	air force
M	stock broker	A	commercial artist
M	builder	A	quilt shop owner
H	lawyer	A	photographer
H	public relations	S	archeologist
S	marine biologist	S	veterinarian
O	librarian	A/S	wildlife photographer
O	Renaissance woman (drifter)	S	astronaut
S	chemistry teacher	H	restaurant owner
S	physical therapist	S	landscape architect
S	nurse (3)	S	chemical engineer
Sp	professional golfer	O	world traveler
A	dancer	A	musician
A	singer (2)	A	actress

They made a graph showing the results of their data collection.

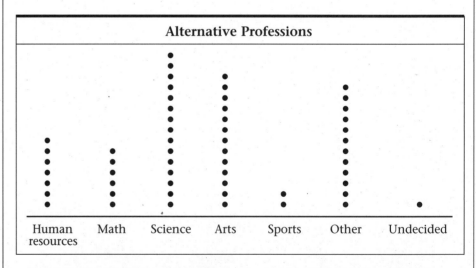

For this project, the distinction between categorical and numerical data can be discussed. While the team did not indicate the mode of the data, if we want to talk about what is typical the only relevant statistic is the mode. Of course, discussing how the categories were chosen and how they were applied is of value as well. Could the team write instructions for coding so that another team could code the list of professions in the same way this team did?

Categories of Years in Education (numerical data)

Another team asked the question, How many years have you been in education? They made a line plot of the data they collected.

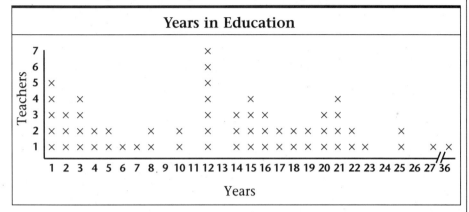

Because these data are numerical, the median and the mode are useful statistics. Range can be used as a measure of spread. Participants may also want to discuss what "been in education" means to clarify what data were collected.

Categories of Desired Age (numerical data)

Another team asked the question, What is your most desired age? Their line plot shows the numerical data grouped by interval.

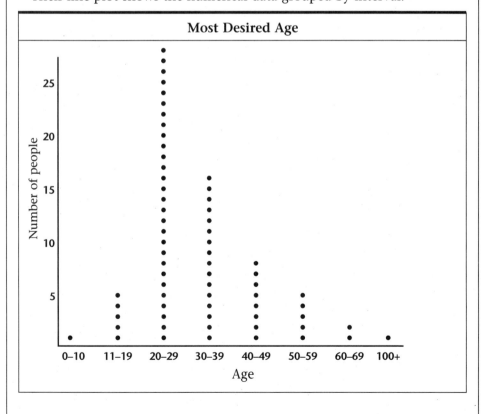

Again, mode, median, and range are appropriate statistics. The mode is really a *modal interval* (20–29 years). How could participants find the median of these data? What would they report if they had only the graph to use to find the median?

It is interesting to note that in the graphs made for the second and third examples, the teams each labeled both the horizontal and vertical axes. With line plots or variations of them, the vertical axis is not needed. You could discuss this observation with participants.

Summary

The mode is used to describe categorical data; the median is used to describe numerical data. You may want to discuss days of the week or months of the year as examples of categorical data that seem to have a natural order (for example, January, February, March) but are not numerical data and therefore cannot have a median. In addition, students may be unclear about what the mode represents (see Handout 16.1).

Confusion with respect to the use of the range may surface. The range is a statistic for numerical data. However, children and adults alike are likely to try to compute a range for categorical data that relates to the frequency count. This sometimes translates to a kind of "popularity range," which is really the range of the number of responses to a set of categorical data.

For example, consider a display of the birth months for a class of students. The data are categorical, and a line graph is organized by months of the year with the number of marks showing the frequency or count of the number of students who have a birthday in that month. You may encounter this kind of summary: The mode birthday month is April and the range (of people who have a birthday in any one month) is 0 to 5. This latter statistic is really a reorientation of the collected data, with the analysis being of how many responses are given for each month (the range of this data being from 0 to 5 people who indicate a given month as their birth month).

If you are using *About Us* as a workshop project, remind participants that they must find time to complete this project before the workshop ends. Their job as teams is to reconsider their questions from the first investigation—they may even change them—and to go through the process of statistical investigation, preparing a final presentation for the last day of the workshop.

Categorical Data, Mode, and Range Problem Sheet

1. Here are data that were collected in response to the question, What's the typical month for people to have their birthdays?

   Months People Have Birthdays

			×								
			×								
			×			×		×			
×			×	×		×		×			
×			×	×	×	×		×	×		
Jan	Feb	Mar	Apr	May	Jun	Jul	Aug	Sep	Oct	Nov	Dec

   Month

   a. Which statement below reflects an understanding of the concept of mode?

   "Most of the people in the class have birthdays in April."

   "April is the month with the most number of people having birthdays."

   b. How might you respond to the statement that reflects a misunderstanding of the mode to help clarify the students' understanding?

2. A student notes, "We can also find the range of birthdays. It is 0 to 5."

   Number of Responses for Each Birthday Month

×						
×						
×						
×	×	×	×			
×	×	×	×		×	
0	1	2	3	4	5	6

   Responses

   a. Can there be a range for categorical data?

   b. Does this reorientation of the data to answer the question "What is the count of responses given for the months of the year?" clarify the student's thinking? Explain why or why not.

INVESTIGATION 17
Shape of the Data: Using Stem Plots

Overview

Participants explore two problems in order to collect data and display it using stem-and-leaf plots, or stem plots. The first problem motivates the need for a way to display data to accommodate greater spread. The second problem shows how data previously collected can be displayed in a different way. The two problems are

- How long can you hold your breath?

- How can we use a stem-and-leaf plot to show the data of dates on pennies from Investigation 6? How does this display compare with the line plots we used earlier?

Assumptions

Participants have done some introductory work with data analysis. In particular, they are familiar with representing data in the form of line plots and are comfortable discussing the concept of the shape of the data as it relates to specific investigations.

Goals

Participants further explore the concept of the shape of the data. This includes a consideration of five components:

- symmetry or skewness

- presence or absence of single or multiple peaks

- center(s) of the data

- degree of spread around the center

- deviations from the regular pattern in the data, including gaps and outliers

As an informal exercise, we begin by talking about clumps, bumps, holes, and what's typical about the data.

Materials

Pens, paper, stopwatches or a clock with a second hand, chalkboard or large chart paper, calculators

Transparency 17.a

Handout 17.1

The stem-and-leaf plot, or stem plot, is introduced as a way to display data. Participants consider some of the common shapes of stem plot distributions: mound-shaped, U-shaped, J-shaped, and rectangular-shaped distributions. Participants compare the stem plot and the line plot as tools for displaying data.

References

Russell, S. J., and R. B. Corwin. *Statistics: The Shape of the Data.* Palo Alto, California: Dale Seymour Publications, 1989.

Landwehr, J. M., and A. E. Watkins. *Exploring Data.* Rev. ed. Palo Alto, California: Dale Seymour Publications, 1995.

Fey, J., W. Fitzgerald, S. Friel, G. Lappan, and E. Phillips. *Data About Us.* Palo Alto, California: Dale Seymour Publications. Forthcoming.

Developing the Activity: Part 1

Participants will decide how to collect data to answer a question about breath holding, then collect the data and organize it using a stem plot.

Pose the Question

Ask participants to consider the following question:

What's the typical amount of time a person in this group can hold his or her breath?

Explain that today they will plan the investigation, and in the next session they will carry it out.

Have participants begin to think about the best way to collect data.

Have each person in your group hold his or her breath one time. Time the duration of each person's attempt, and record the time. Then, as a group, think about how you should plan and implement your experiment.

Once groups are ready, continue the discussion.

Are there different ways we could collect these data? Should each of you hold your breath once, or several times? Should we then pull the longest times from the raw data? What other problems might we run into while collecting the data?

Have participants make estimates about how long they think they can hold their breath and what might be typical for the group.

Collect the Data

Participants will need to discuss how they will carry out the experiment. Do they want to time themselves, or is it better to work in pairs? If someone starts to giggle, does he or she get another chance? Do we want to have more than one try and record the longest time? Do we collect the data in seconds, or in minutes and seconds?

You and participants will want to make sure you agree on the procedure that will be used so that everyone goes about the data-collection process in a similar fashion.

Analyze the Data

With the participants, begin the analysis by attempting to make a line plot to display the data.

Let's try putting our data on a line plot. What was the shortest amount of time someone held his or her breath?

What was the longest amount of time someone held his or her breath?

This gives us a range of ____ to ____ seconds.

On the board or chart paper, begin to number consecutively in the small intervals (for example, by ones, twos, or fives) that you and participants agree are needed to make a line plot. Display the participants' data on the line plot.

What can we say about this preliminary data? What does it tell us about the people in this room?

Have a brief discussion about participants' observations.

The range is ____ to ____ seconds, and if we use a line plot, the data are quite spread out. It's hard to see whether there are any real clumps. Let's display the data using another distribution: a stem-and-leaf plot, or stem plot.

Demonstrate how to make a stem plot with the participants' data. First make the stem plot, just recording the times as individuals call them out. Then demonstrate how to reorder the data, making a second version of the stem plot with ordered data.

Here is an example of a collection of data showing how long 21 teachers could hold their breath (in seconds).

56 43 31 22 91 48 79 38 32 59 78 45
35 62 55 39 46 42 80 47 79

To make a stem plot of the data, we divide each value into tens and units. The tens become the "stem" of the plot, and the units become the "leaves."

```
2 | 2
3 | 1 8 2 5 9
4 | 3 8 5 6 2 7
5 | 6 9 5
6 | 2
7 | 9 8 9
8 | 0
9 | 1
```

The top line illustrates one data point in the twenties: 22. The second line shows five data points in the thirties: 31, 38, 32, 35, 39.

Once we have quickly organized the data into a stem plot, we can rearrange the leaves for each tens place in order.

```
2 | 2
3 | 1 2 5 8 9
4 | 2 3 5 6 7 8
5 | 5 6 9
6 | 2
7 | 8 9 9
8 | 0
9 | 1
```

In the ordered stem plot, many features of the data are now apparent. The data range is from 22–91 seconds. A large cluster of data—about half of the data—is in the thirties and forties. If we include the fifties in that cluster, we account for two thirds of the data. Data points above 60 seconds are scattered; however, an interesting small cluster occurs at 78–80 seconds.

Now that we have displayed the data in a stem plot, what can you see about the shape of the data? How are line plots and stem plots alike? How are they different?

Interpret the Results

Have participants work in groups to discuss and record their descriptions of the data, including observations about where the data are clumped, if there is more than a single clump, and where there are holes or outliers. Then have them decide their group's typical time for breath holding.

Participants may report their findings and provide reasoned discussions for the decisions they have made.

Developing the Activity: Part 2

Participants will make a stem plot of the pennies data used or collected in Investigation 6, and compare what the stem plots and line plots reveal about the data.

Pose the Question

Remind participants of the work they did with the pennies data in Investigation 6. Have them respond to this question:

What would our pennies data look like if we represented them using a stem plot in addition to a line plot?

Collect the Data

Pass out Handout 17.1, which displays the pennies data in Distributions *A* and *B*.

Analyze the Data

Discuss with participants dropping the "19" from each year and then representing the remaining two digits of the years as tens and ones. Some participants may notice that the data are already ordered because they are displayed on a line plot. Have participants make a stem plot of the data in Distribution A.

```
6 | 7 9
7 | 4 6 6 6 6 7 7 7 7 7 9
8 | 0 1 1 2 2 3 3 3 3 3 3 4 4 4 4 4 4 4 5 5 5 5 5 5 5 5 6 6 6 6 6 7 7 8 8 8 8 9 9 9 9 9 9
9 | 0 0 1 1 1 1 2 2 2 2 2 2 2 3 3 3 3 4 4 4 4 4 4 4 4 4 4 4 4 4
```

Discuss the shape of the data from Distribution A.

Next, demonstrate how to make a stem plot marked in intervals of 5.

```
6 |
• | 6 7 9
7 | 4
• | 6 6 6 6 7 7 7 7 7 9
8 | 0 1 1 2 2 3 3 3 3 3 3 4 4 4 4 4 4 4 4
• | 5 5 5 5 5 5 5 6 6 6 6 6 7 7 8 8 8 8 9 9 9 9 9 9
9 | 0 0 1 1 1 1 2 2 2 2 2 2 3 3 3 3 4 4 4 4 4 4 4 4 4 4 4 4 4
```

Again, discuss the shape of the data, commenting on the different graphical representations of the data in the two stem plots.

Have participants do a similar activity with Distribution B. They may want to make comparisons between the two stem plots.

Take a few minutes to discuss the common types of distributions we might see in a stem plot: mound-shaped, U-shaped, J-shaped, and rectangular-shaped distributions. The stem plot shows the shape of the data—their distribution—in some common ways (see Transparency 17.a). The mound-shaped curve (usually called a *bell curve*) is the most common. The U-shaped distribution suggests that there may be two underlying groups—each of which is a mound-shaped distribution—corresponding to the two peaks. The J-shaped and rectangular-shaped curves occur less frequently. In either of these cases, you will want to try to understand the data better: are there limits or boundaries to possible values? If so, what do these limits or boundaries mean? These ways of describing the shape of the data provide some common language.

Interpret the Results

Have participants consider the line plots and stem plots of the pennies data, what each type of plot shows, and when they might want to use each.

Discuss the concept of using intervals to group frequencies of data, which is clearly illustrated by the stem plot. Grouping by intervals is an important and often-used concept in statistics.

Shapes of Stem-and-Leaf Plots

Mound-shaped Distribution

```
1 |
2 | 2
3 | 1 2 5
4 | 2 3 5 6 7 8
5 | 5 6 9 4 5
6 | 2 3 7
7 | 8 9
8 | 0
```

U-shaped Distribution

```
1 |
2 | 2 4
3 | 1 2 5 8 8
4 | 2 3 5
5 | 5 6
6 | 2 3 7 7 7 9 9
7 | 8 9 9 9
8 |
```

J-shaped Distribution

```
1 |
2 | 2
3 | 1 2
4 | 2 3 5
5 | 5 6 9 4 5
6 | 2 3 6 6 7 7 7
7 | 0 0 3 3 3 7 9 9
8 |
```

Rectangular-shaped Distribution

```
1 |
2 | 2 2 3 7
3 | 1 2 5 9
4 | 2 3 5 6 7
5 | 5 6 9 4 5
6 | 2 3 7 7
7 | 0 0 2 8 9
8 |
```

Reference

Landwehr, J. M., and A. E. Watkins. *Exploring Data.* Rev. ed. Palo Alto, California: Dale Seymour Publications, 1995.

HANDOUT 17.1

Distribution A

Dates of Pennies

```
                              x x x x x x x x x x x x x x x
                                              x x x x
                                          x x x x x x
                                            x x x x
                                              x x
                                          x x x x x x
                                            x x x x
                                              x x
                                          x x x x x
                                        x x x x x x x
                                        x x x x x x x
                                        x x x x x x
                                            x x x
                                          x x x
                  x x x x x
                  x x x x
            x
  x x
  x
  66 67 68 69 70 71 72 73 74 75 76 77 78 79 80 81 82 83 84 85 86 87 88 89 90 91 92 93 94
```
Year (19--)

Distribution B

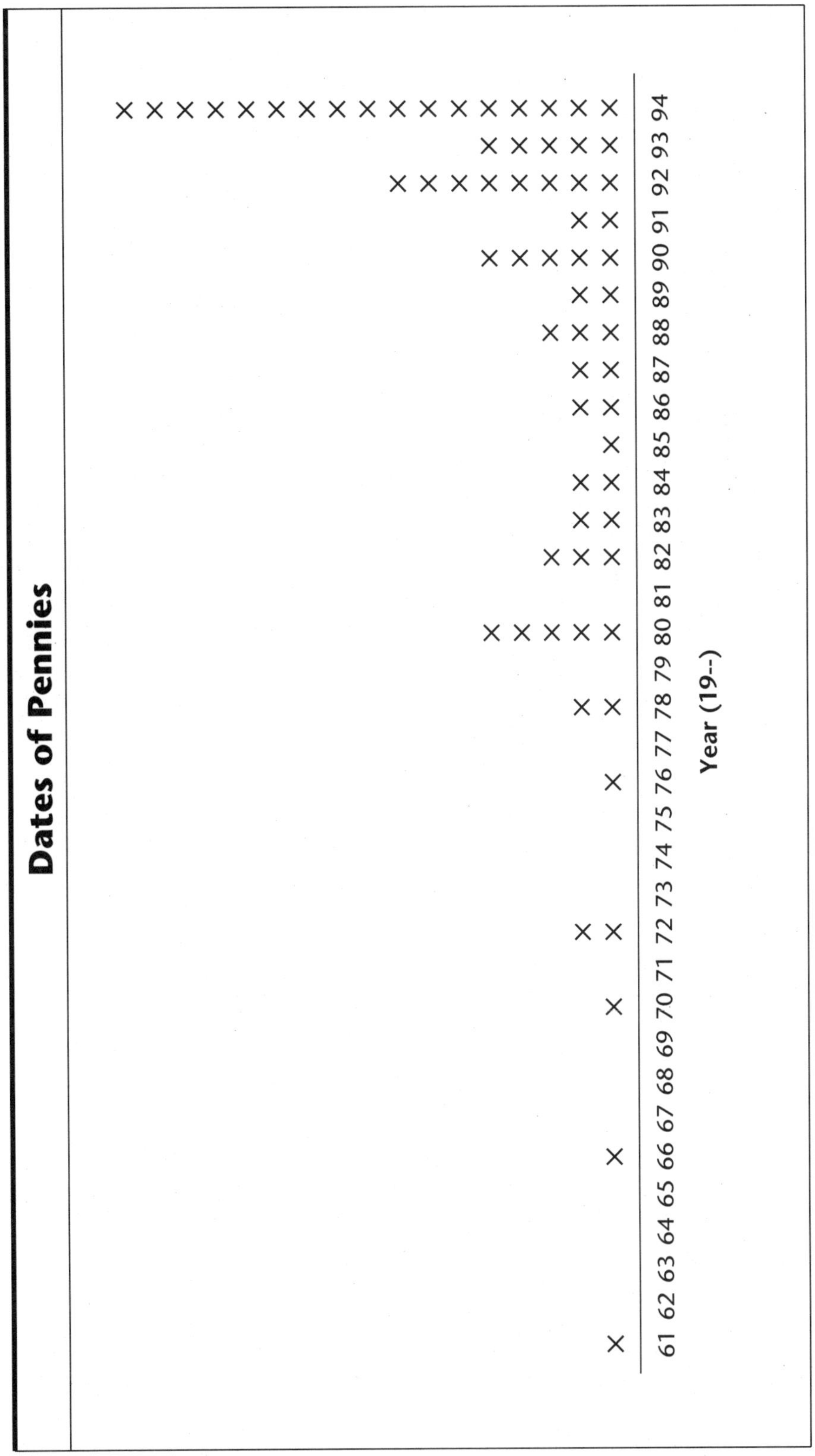

INVESTIGATION 18
Shape of the Data: Stem Plots to Histograms

Overview

Participants explore ways to represent data that are grouped in intervals, and they use their knowledge of stem-and-leaf plots to construct an understanding of histograms. Participants consider three problems. The first problem introduces the use of histograms by linking the development to work done with stem plots. In the second problem, participants make histograms using data sets they explored in earlier investigations. In the third problem, they consider data represented in a histogram, deciding the best way to group the data to make a point. The problems concern these questions:

- What is the typical time it takes a group of students to travel to school?

- How can we show the breath-holding data and the pennies data using histograms?

- Given data on students' allowances, what is the typical allowance?

Assumptions

Participants have completed work related to the use of stem plots. They will be comfortable with grouping data in fives and tens using this representation.

Goals

Participants will be able to create, read, and interpret histograms to represent data. In particular, they

- link the structure of stem plots with the structure of histograms, recognizing the ways they are similar and different for representing grouped data

- make histograms from data represented in stem plots

- explore ways to group data using histograms, and evaluate the impact of the grouping method on the interpretation of the data

Materials

Copies of stem plots from Investigation 17 showing the breath-holding data and the pennies data, calculators

Transparencies 18.a through 18.h

Handout 18.1

Developing the Activity: Part 1

Participants explore a stem plot of data about students' travel times to school, the related frequency table, and the construction of histograms from the information in the stem plot and frequency table.

Pose the Question

Here is a problem that an elementary class decided to investigate.

The students were interested in how they used their time. They brainstormed a list of ways—such as sleeping, eating, and after-school sports. James reminded them that some of their time was used traveling back and forth to school. Some of the students thought this shouldn't count, because it really wasn't much time. Others disagreed. The class wondered: What is the typical time it takes to travel to school?

Here is the stem plot the class made to show data they collected about the number of minutes it took them to travel to school on one day (see Transparency 18.a and Handout 18.1).

Minutes to Travel to School

```
0 | 3 3 5 7 8 9
1 | 0 2 3 5 6 6 8 9
2 | 0 1 3 3 3 5 5 8 8     2 | 3  means 23 minutes
3 | 0 5
4 | 5
```

Looking at this stem plot, how could you describe the data? What do you know about the time it takes for students to travel to school? What questions do you have concerning these data?

How might we answer their question: What is the typical time it takes a student to travel to school?

Collect the Data

Give participants time to discuss these questions. You may want them to talk about the questions posed on Handout 18.1 as well.

Refocus participants' attention on the stem plot. Ask them to consider the following questions:

What do the 0, 1, 2, and so on in the stem mean? What do these numbers tell you about the data that are in the leaves?

Emphasize that the 0 refers to the interval of 0–9 minutes in the data. Similarly, the 1 refers to the interval of 10–19 minutes.

We can make a frequency table to show the intervals of the data displayed in the stem plot (see Transparency 18.b).

Minutes to Travel to School	
Minutes	Number of people
0–9	6
10–19	8
20–29	9
30–39	2
40–49	1

It is important to emphasize that the 0, 1, and so on have place value and indicate an interval of data values.

Analyze the Data and Interpret the Results

Next, have participants explore how to make a histogram. Using Transparency 18.c, ask them to discuss ways in which the histogram (a new word) is related to the stem plot.

Describe how the information in the stem plot and its related frequency table of intervals is used to make this histogram.

Discuss the various components of the histogram:

- the two axes and the plot elements

- blocks that *touch* and why they touch

- how the components relate to data in the stem plot and in the table of data

Here are two new data values: Kyle took 46 minutes to travel to school, and Janette took 15 minutes. How would we add these data to the stem plot, the table, and the histogram?

This question is designed to focus participants' attention on the "end points" of the intervals.

Where are times that are multiples of 5 minutes actually recorded—in what intervals do they "fall"? What about times that are multiples of 10 minutes?

Teacher Notes

Histogram: A block graph that shows continuous data that have been organized using frequency intervals

Once participants are comfortable with this discussion, continue, using Transparency 18.d. This introduces the use of intervals of 5-minutes width.

Describe how the information in the stem plot and its related frequency table is used to make this histogram.

Add the two new data values to the histogram: Kyle took 46 minutes to travel to school, and Janette took 15 minutes.

Again, this focuses participants' attention on the "end points" of an interval.

Where are times that are multiples of 5 minutes actually recorded—in what intervals do they fall? What about times that are multiples of 10 minutes?

Transparency 18.e shows the two histograms (with intervals of 10 and intervals of 5) displayed together.

Let's compare the two histograms, looking at the shape of the data in each. Discuss what happens to the shape of the data when the data are grouped in intervals of different sizes.

The detail in the data gradually "disappears" as the intervals increase in width. Depending on the purposes of the analysis, either representation may be useful.

Developing the Activity: Part 2

Work with participants to make histograms of the data collected or used earlier about breath-holding capacity and dates on pennies. Encourage them to experiment with different intervals for grouping the data, including intervals that are not multiples of 5.

Developing the Activity: Part 3

Participants examine data in a histogram and explore the best way to group the data to make a particular point.

Pose the Question

Display Transparency 18.f. This histogram shows allowances for 60 students. Students could use data such as these to make an argument for a particular allowance.

The first bar shows that five students received an allowance less than $0.50; one student might have received no allowance, another student might have received $0.35, but no student in this group received an allowance of $0.50.

The second bar shows that three students received an allowance of at least $0.50 but less than $1.00; one student might have received an allowance of $0.50 or $0.75, but no student received an allowance of $1.00.

Collect the Data

Provide participants with the section of Handout 18.1 that shows this bar graph and presents the problem.

Analyze the Data and Interpret the Results

The problem presented on Handout 18.1 can be explored by small groups of participants.

Groups will want to experiment with making different histograms. Two examples are provided as Transparencies 18.g and 18.h, which you may want to use in the discussion following the group work.

Again, you will want to discuss what happens to the shape of the data when the data are grouped in different-size intervals. The detail in the data gradually disappears as the width of the intervals increases. Depending on the purposes of the analysis, either representation may be useful.

Summary

Histograms are actually area graphs. The horizontal axis functions as a continuous scale, with initial and end points of each "bar block" noted as the scale markers. It is possible to have unequal column widths, depending on the sizes of the corresponding intervals. For our purposes, intervals are derived from stem plots and are equal, meaning the widths of the blocks are equal. The frequency scale on the vertical axis represents the individual data items in the stem plot by highlighting the count of those data items.

Minutes to Travel to School: Stem-and-Leaf Plot

```
0 | 3 3 5 7 8 9
1 | 0 2 3 5 6 6 8 9
2 | 0 1 3 3 3 5 5 8 8
3 | 0 5
4 | 5
```

2 | 3 means 23 minutes

Minutes to Travel to School: Frequency Table

Minutes	Number of people
0–9	6
10–19	8
20–29	9
30–39	2
40–49	1

Minutes to Travel to School: Histogram (10 min. intervals)

Minutes to Travel to School	
0	3 3 5 7 8 9
1	0 2 3 5 6 6 8 9
2	0 1 3 3 3 5 5 8 8
3	0 5
4	5

Minutes to Travel to School	
Minutes	Number of people
0–9	6
10–19	8
20–29	9
30–39	2
40–49	1

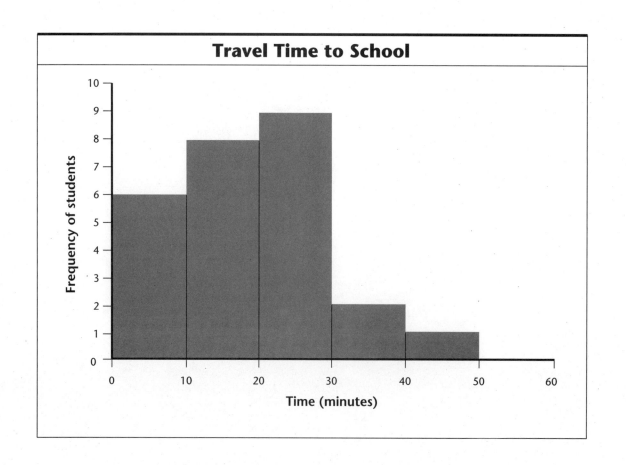

Minutes to Travel to School: Histogram (5 min. intervals)

Minutes to Travel to School	
0	3 3
•	5 7 8 9
1	0 2 3
•	5 6 6 8 9
2	0 1 3 3 3
•	5 5 8 8
3	0
•	5
4	
•	5

Minutes to Travel to School	
Minutes	Number of people
0–4	2
5–9	4
10–14	3
15–19	5
20–24	5
25–29	4
30–34	1
35–39	1
40–44	0
45–49	1

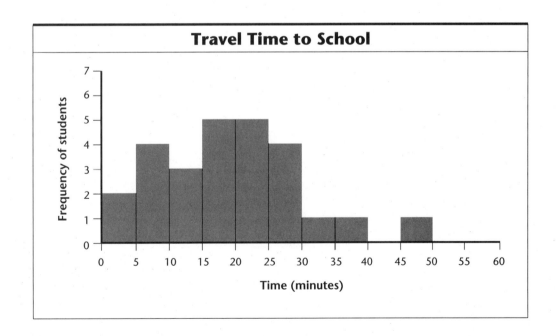

Travel Time to School

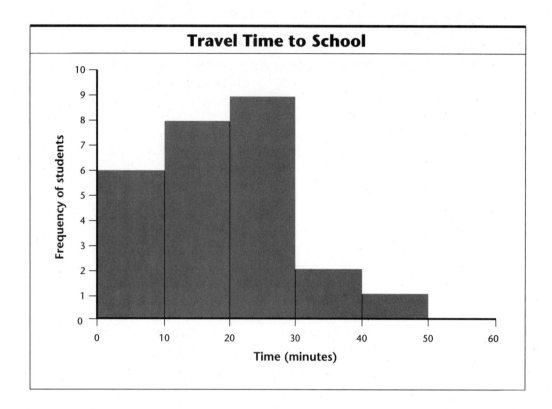

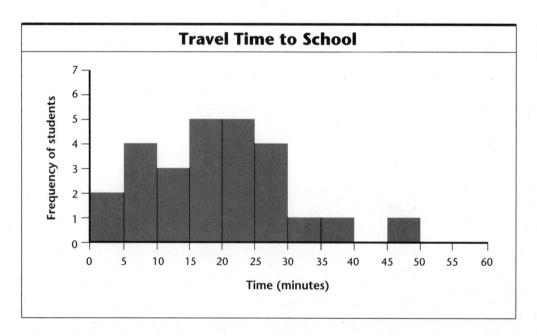

Allowances of 60 Students: Histogram ($.50 intervals)

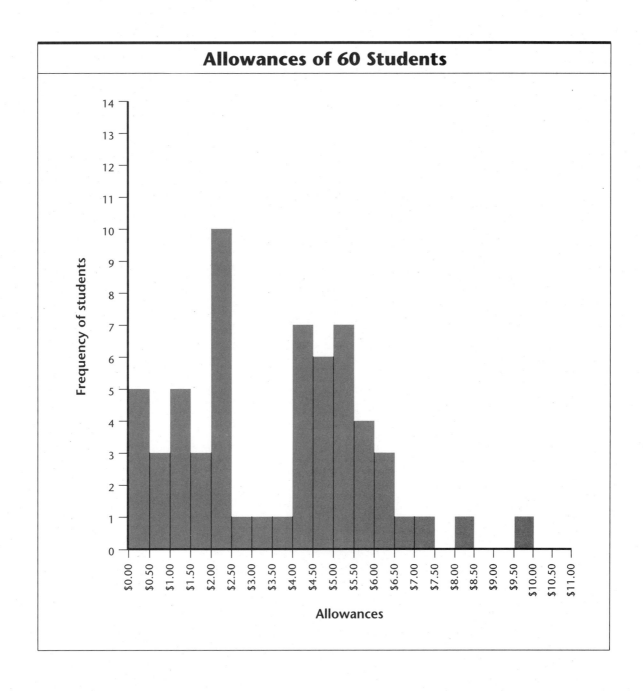

Allowances of 60 Students: Histogram ($1.50 intervals)

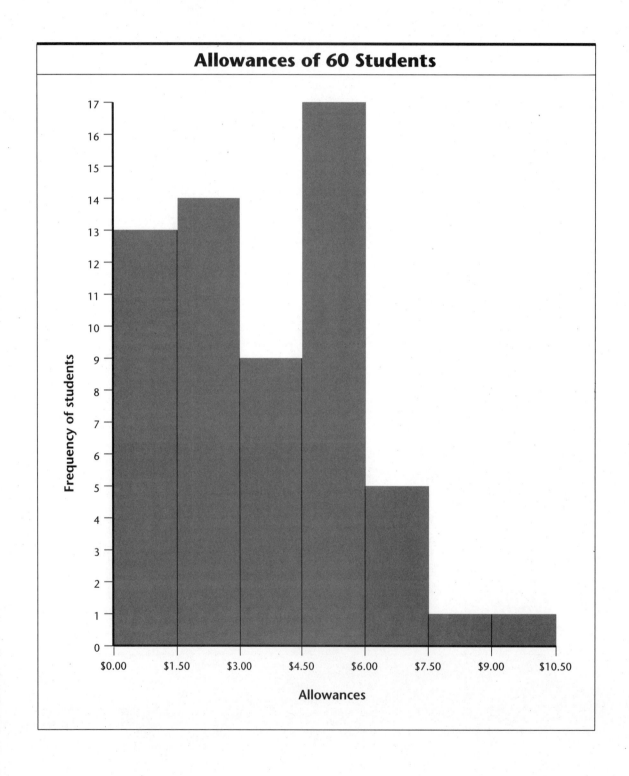

Allowances of 60 Students: Histogram ($1.00 intervals)

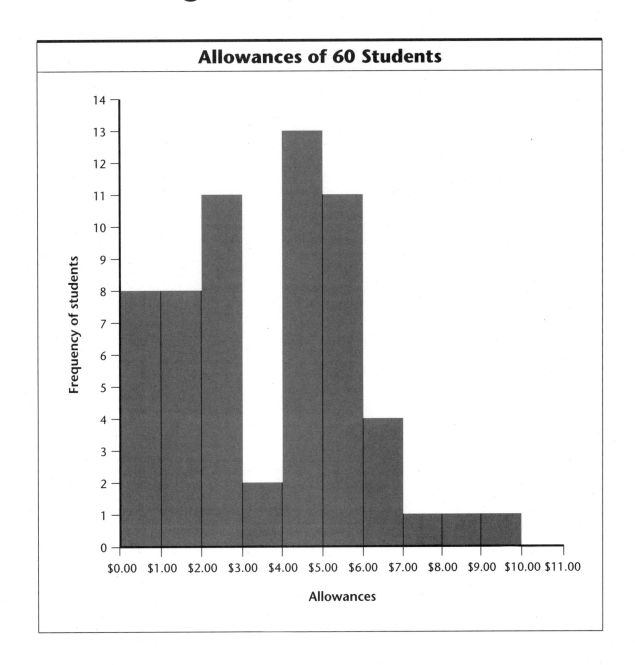

HANDOUT 18.1

Stem Plot to Histogram Problem Sheet

1. Students were interested in how they used their time. They brainstormed a list of ways such as sleeping, eating, and after-school sports. James reminded them that some of their time is used just traveling back and forth to school. Some of the students thought this shouldn't count because it really wasn't much time at all. Others disagreed. The class wondered, *What is the typical time it takes to travel to school?*

   Here is the stem plot they made to display data they collected about the number of minutes it took them to travel to school in one day.

Minutes to Travel to School	
0	3 3 5 7 8 9
1	0 2 3 5 6 6 8 9
2	0 1 3 3 3 5 5 8 8
3	0 5
4	5

   2 | 3 means 23 minutes

   a. Using the stem plot, describe what you know about how long it takes these students to travel to school.

   b. How many students took less than 10 minutes to travel to school? How can you tell?

   c. Can you use this graph to answer this question: How many students took *at least* 15 minutes or more to travel to school? Explain why or why not.

d. How many students are in the class? How can you tell?

e. What is the typical time it takes for students to travel to school? Explain your answer.

f. Make a histogram to show the information displayed in the stem plot.

2. Here is a different version of the original stem plot.

\multicolumn{2}{c}{**Minutes to Travel to School**}	
0	3 3
•	5 7 8 9
1	0 2 3
•	5 6 6 8 9
2	0 1 3 3 3
•	5 5 8 8
3	0
•	5
4	5

2|3 means 23 minutes

a. How are the data grouped in this stem plot?

b. Make a histogram to show the information displayed in the stem plot.

HANDOUT 18.1

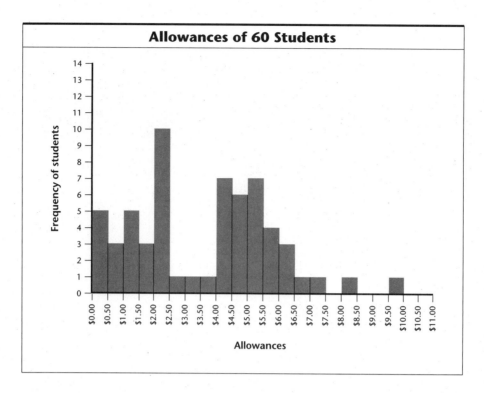

3. The histogram above shows allowances for 60 students. The first bar shows that five students received an allowance of less than $0.50, meaning that one student might receive no allowance, another student might receive $0.35, but no student received $0.50. The second bar shows that three students received an allowance of at least $0.50 but less than $1.00, meaning that one student might receive $0.50 or $0.75 but no student received $1.00.

   a. What do you know about the allowances of students in the two tallest bars? Explain.

   b. One student's parents have agreed to look at these data and then decide what allowance to give her. Using the histogram, what allowance do you think the student should make a case for? Why?

   c. What happens if you group the data by intervals greater than $0.50? Given the data in the original graph, what intervals can you use?

   d. Make one or more histograms showing the data regrouped in different intervals. Does regrouping change the message of the graph—that is, can you argue more effectively for a specific allowance?

INVESTIGATION 19

Curriculum Overview

Overview

The activities presented during the workshop are designed to help participants learn about statistics and data analysis and about how to teach these ideas to their students. It is important that participants have the opportunity to think about and to discuss where teaching statistics and data analysis fit into their mathematics curriculum.

Assumptions

Participants use their states' course of study and their districts' curriculum standards to guide discussion during classroom activities.

Goals

Participants will

- identify key statistical concepts and relate them to their own programs of mathematics instruction

References

Curriculum and Evaluation Standards for School Mathematics. Reston, Virginia: National Council of Teachers of Mathematics, 1989.

Curriculum and Evaluation Standards for School Mathematics Grades K–6 Addenda Series. Ser. ed. Meriam A. Leiva. Reston, Virginia: National Council of Teachers of Mathematics, 1992.

Developing the Activity

Divide participants into groups by grade level. Have them review their state and local curriculum guidelines, looking for links with statistics and data analysis.

The instructor's role is to ask questions and encourage discussion as needed. Questions, issues, and ideas for the classroom should be recorded by one person in the group for later discussion.

Materials

Curriculum guides; local, state, and national textbooks; resource books for teachers; copies of the Teach-Stat concept map

Using curriculum guides and textbooks, discuss the place for statistics and data analysis in the elementary-grades mathematics curriculum. Guide discussion of the following questions:

What concepts and processes are appropriate for each grade level?

In particular, consider the concepts and processes shown in the Teach-Stat concept map.

How has the curriculum changed with respect to statistics since you began teaching?

How much time do you usually spend teaching statistics?

How can ideas related to statistics and data analysis be incorporated into other subjects?

Ask for some examples. Make sure these ideas are recorded.

What types of graphs are appropriate for student use at each grade level? What methods of display do students choose to use?

Help participants realize that children's understanding of graphs begins early and develops throughout the elementary grades.

What types of activities are appropriate for students at each grade level?

Focus on the idea that activities may vary in length from a one-period activity to a year-long activity. Encourage participants to think about activities they have worked on during their participation in the workshop.

How would you modify your experiences for use in your classes?

Remember, the activities will not be appropriate for every teacher to use in every class.

Ask participants to reflect on the activities they completed this week.

Which activities might be appropriate for your classes?

How would you modify them for your grade level?

Do you have any ideas for other activities?

INVESTIGATION 20
Shape of the Data: Problem-Solving

Overview

Participants have an opportunity to apply material with which they have worked previously. Specifically, they work with line plots, bar graphs, stem-and-leaf plots, and histograms.

Assumptions

Participants know how to read and make line plots, bar graphs, stem plots, and histograms.

Developing the Activity

Have participants complete Handout 20.1 in small groups. As you circulate and work with individual groups, identify areas that should be discussed with the entire group of participants.

As a whole group, talk about any issues that arose during the group work and any other questions that participants want addressed.

Reference

Fey, J., W. Fitzgerald, S. Friel, G. Lappan, and E. Phillips. *Data About Us*. Palo Alto, California: Dale Seymour Publications. Forthcoming.

Materials

Calculators

Handout 20.1

Problem Solving 173

HANDOUT 20.1

Shape of the Data Problem Sheet

Part 1

Identify the median and range for each distribution below.

Length of First Names

```
        ×
        ×  ×
        ×  ×  ×
        ×  ×  ×  ×
1  2  3  4  5  6  7  8  9  10 11 12 13 14
            Number of letters
```

Length of First Names

```
      ×  ×        ×
      ×  ×  ×     ×  ×  ×        ×
1  2  3  4  5  6  7  8  9  10 11 12 13 14
            Number of letters
```

Length of First Names

```
                  ×
                  ×
                  ×
                  ×
            ×     ×
            ×     ×
            ×  ×  ×  ×
         ×  ×  ×  ×  ×  ×
         ×  ×  ×  ×  ×  ×  ×
1  2  3  4  5  6  7  8  9  10 11 12 13 14
            Number of letters
```

174

Part 2

1. Using the bar graph below, answer the following questions.

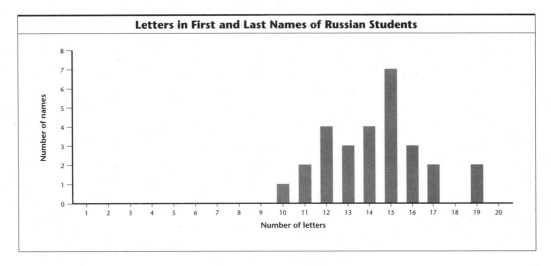

   a. What information does the tallest bar give you? Explain your reasoning.

   b. How many students are in the middle-grades class in Russia? Explain.

   c. What is the range of the number of letters in students' names in this class? Explain.

   d. What is the median number of letters in students' names? Explain.

   e. Marina Arkhangelskaja and Olga Aladeeva were absent the day name lengths were graphed. Show how to add their data to the bar graph.

2. How is the bar graph below similar to the first bar graph? Different from the first bar graph? Explain your reasoning.

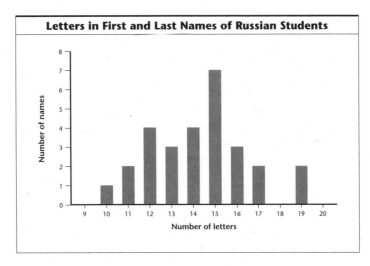

HANDOUT 20.1

Part 3

One class investigated the problem of how many paces it took to travel from their classroom to the library. They measured the distance by counting the number of paces they walked; every step made on the right foot counted as one pace. They compared their paces for the same route from their classroom to the library and found the following results.

```
              Paces from Classroom to Library
                                    ×
                                    ×
                                    ×      ×
                                    ×      ×
                       ×            ×      ×      ×
              ×        ×    ×   ×   ×      ×      ×      ×
              ×   ×    ×    ×   ×   ×      ×      ×      ×
             17  18   19   20  21  22     23     24     25
                            Number of paces
```

1. What is the median number of paces needed to travel the distance?

2. Make a bar graph that shows this information. Explain how the bar graph is similar to and different from the line plot.

3. Who has the shorter pace: the student who traveled the distance in 17 paces, or the student who traveled the distance in 25 paces? Explain your reasoning.

Part 4

1. Make a line plot showing the lengths of 12 names so that the median length is 6 letters.

2. Make a line plot showing the lengths of 15 names so that the median length is 7 letters and the range of lengths is 3–10 letters.

3. Make a line plot showing the lengths of 20 names so that the median length is 5½ letters and the range covers a spread of 8 lengths.

Part 5

The information below was collected from a large number of middle-grades students. It is shown as a *table of data*. The first column shows the kinds of pets the students had, and the second column shows how many of each kind of pet they had. You can see that 61 of the total of 1000 pets are birds and 184 of the total of 1000 pets are cats.

Pets Students Have	
Kind of pet	Count of kind of pet
bird	61
cat	184
chicken	29
cow	9
crab	9
dog	180
duck	9
fish	303
frog	12
gerbil	17
goat	4
guinea pig	12
hamster	32
horse	28
lizard	10
mouse	10
pig	6
rabbit	42
rat	4
snail	12
snake	9
turtle	13
worm	8

From this data, we cannot tell how many students were surveyed. We only know that 1000 pets were counted.

1. What might be a good estimate of how many students were surveyed? Explain your reasoning.

2. Make a bar graph to display this categorical data. Think carefully about how you will label the horizontal and vertical scales.

3. Write a paragraph about what you know about pets students have based on the information in the table.

HANDOUT 20.1

Part 6

Look at the following table of data about typical pulse rates for a number of different kinds of animals, including humans. From the table, we can see than an alligator has a pulse rate of 47 beats per minute and a bass has a pulse rate of 128 beats per minute.

Pulse Rates for Animals

Animal	Pulse rate	Animal	Pulse rate
alligator	47	groundhog	80
bass	128	guinea pig	270
basset hound	92	haddock	37
beagle	120	horse	35
bear	55	human adult	70
beaver	140	human baby	148
Boston terrier	120	Irish terrier	100
camel	25	kangaroo	120
carp	100	lion	42
cat	55	mule	40
cod	48	ostrich	70
collie	100	Pekinese	125
cow	55	perch	59
crocodile	48	pig	70
deer	70	pigeon	170
dolphin	110	pointer	90
donkey	40	porcupine	300
duck	268	rabbit	150
elephant	30	salmon	38
fox	240	seal	44
fox terrier	95	shark	29
frog	48	sheep	75
giraffe	66	trout	35
goat	80	turkey	211
goldfish	38	whale	16

1. Make a stem plot of this data. Let the stems be 1, 2, 3, 4, 5, 6, 7, 8, 9, 10, 11, and so on. Let the leaves be the unit digit in each pulse rate. Write the names of the animals to the side as you record them on the final version of your stem plot. For example:

   ```
   1 | 6               whale
   2 | 5 9             camel, shark
   3 | 0 5 5 7 8 8     elephant, trout, horse, haddock, goldfish, salmon
   4 |
   ```

2. Describe the shape of the data.

3. Find the median pulse rate for the animals shown. What animals are close to or at the median pulse rate?

4. What theories do you have about why animals have the pulse rates they have? Explain your theories.

Reference

Friel, S., J. Mokros, and S. J. Russell. *Statistics: Middles, Means, and In-Betweens*. Palo Alto, California: Dale Seymour Publications, 1992.

Part 7

The three histograms that follow display the animal pulse-rate data.

1. Explain how the data are grouped in Histogram A.

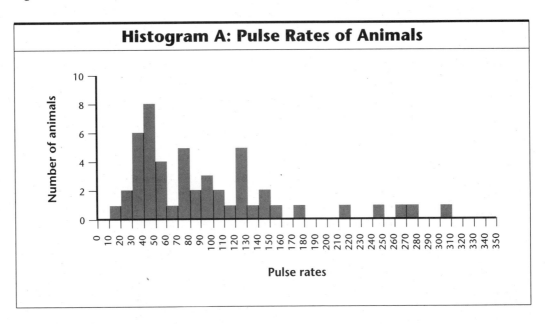

2. Explain how the data are grouped in Histogram B.

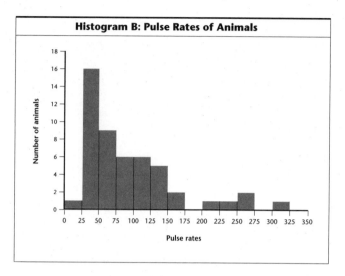

3. Explain how the data are grouped in Histogram C.

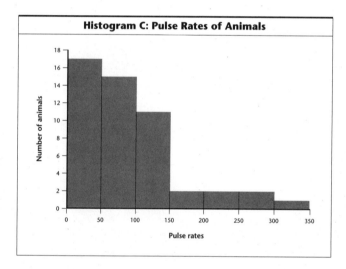

4. Which histogram do you prefer? Explain your preference.

Part 8

A group of middle-grades students were curious about the heights of students at different grade levels, so they gathered data about the heights of students grades 1, 5, and 8. The data they collected are shown on the next page.

1. Make a stem plot that shows the data for each grade.

2. Where do the heights of the grade 1 students cluster? Are there any outliers? Explain your answer.

3. Where do the heights of the grade 5 students cluster? Are there any outliers? Explain your answer.

4. Where do the heights of the grade 8 students cluster? Are there any outliers? Explain your answer.

5. Make a stem plot showing the height data from the three classes together.

6. Make a histogram of the height data for the three classes. Group the data in intervals of 10 centimeters (like the stem plot).

7. Using the histogram, describe the shape of the data.

Height of Students (centimeters)

Grade 1	Grade 5	Grade 8
117	138	147
118	138	156
119	138	159
119	139	160
120	141	160
122	142	161
123	144	162
125	146	162
125	147	162
127	147	162
127	147	163
128	150	164
128	150	165
129	151	165
132	151	168
132	151	168
133	151	168
	152	168
	152	169
	152	171
	152	172
	153	174
	153	176
	155	
	155	
	156	
	156	
	157	
	158	
	171	

INVESTIGATION 21

Graphing: A Minilecture

We live in an information age. Rapidly expanding bodies of knowledge combined with increased uses of technology clearly indicate that we as adults and the children currently enrolled in elementary grades will need to be able to evaluate and use vast amounts of data in personal and job-related decisions.

Skills in gathering, organizing, displaying, and interpreting data are important for students in all content areas. Graphing activities incorporate knowledge and skills from a variety of mathematical topics, integrating geometric ideas with computational skills and classification tasks with numerical understandings.

Graphs provide a means of communicating and classifying data; they allow for the comparison of data and display mathematical relationships that often cannot be easily recognized in numerical form. The traditional forms of graphs are picture graphs, bar graphs, line graphs, and circle graphs. Newer plotting techniques include line plots, stem-and-leaf plots, and box plots.

This minilecture is an overview of the different types of graphs and their use and comprehension.

Traditional Graph Forms

Pictographs

Pictographs (picture graphs) use pictures to depict quantities of objects or people with respect to labeled axes. They are used with discrete (noncontinuous) data. The symbols used must be the same size and shape. They may represent real objects (for example, a stick figure to represent a person, a carton to represent milk drunk by students) or may take the form of something more abstract (such as a triangle or square).

In pictographs without a legend, the symbol and the item it represents are in one-to-one correspondence. When a legend is used, the ratio of each symbol to the number of objects it represents must be taken into consideration when interpreting the graph. Fractional parts of symbols (for example, one half of a picture) may cause some difficulties for children. Data presented in pictographs may also be

Materials

Transparencies 21.a through 21.q

displayed in bar graphs. Converting pictographs to bar graphs is one way to help children move from semiconcrete representations of data to more abstract forms.

(Note: Transparency 21.a is used later.)

Transparencies 21.b, 21.c, and 21.j

Bar Graphs

Used horizontally or vertically, bar graphs compare frequencies of discrete quantities using rectangular bars of uniform width. The heights (or lengths) indicate value or frequency. The bars are constructed within perpendicular axes that intersect at a common reference point, usually zero. The axes are labeled.

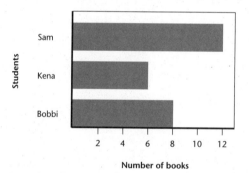

Multiple or double bar graphs are used to compare discrete, stratified data (that is, data collected from particular groups). For example, when asking children to vote for their favorite pet, color, book, or game, the results may be organized according to boys' responses and girls' responses.

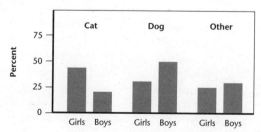

Transparencies 21.d and 21.e

Line Graphs

A line graph is used to show change over time for continuous data. Points are plotted within coordinate axes to represent change over a period of time or any linear functional relationship. The labeled axes intersect at a common point, usually zero. The units of division on each axis are equally spaced, and the graphed points are connected by straight or dotted lines. When children keep a record over a period of time of such data as their height or weight or the daily average temperature, line graphs are appropriate displays.

Multiline graphs are used to compare two or more sets of continuous data—for example, to compare the height or weight of two children over a period of time (such as four months or one year), or the heights of two (or more) plants over a period of time (such as one to two months after planting seeds).

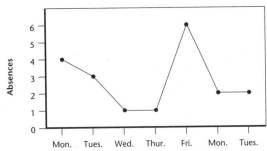

Transparency 21.j

Circle Graphs

The area of the circle graph (pie graph, pie chart, area graph) is divided into sections by lines emanating from the center of the circle. Circle graphs are appropriate representations when children have an understanding of fractions; they provide children with a means of displaying the relationship of parts to whole.

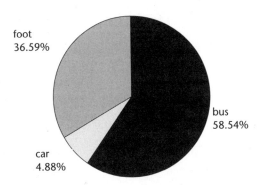

Children may informally create circle graphs before they are able to measure angles and figure proportions. For example, counters representing the total units may be evenly spaced around a circle. A radius is drawn where the divisions occur to divide the circle into appropriate parts. A second informal method is to mark units on a strip and then loop the strip to form a circle, drawing radii as appropriate.

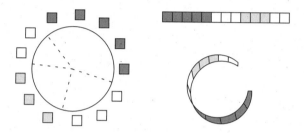

Some New Techniques

Line Plots

Line plots are a quick, simple way to organize data along a number line; the Xs (or other symbols) above a number on a single horizontal axis represent the tally (frequency) for that data value. Unlike a bar graph, in which data may be lost in the grouping, none of the data gets lost in a line plot.

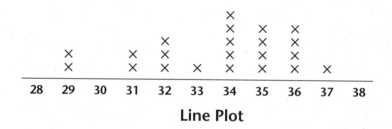

Line Plot

Transparency 21.0

Stem-and-Leaf Plots

Stem-and-leaf plots, or stem plots, are characterized by a "separation" of the digits in numerical data. In a simple stem plot, the tens digits are listed in one column and the ones digit are listed in a row next to the respective tens digit.

Teacher Notes

When rotated 90 degrees counterclockwise, the stem plot resembles a bar graph. The development of the stem plot and the box-and-whiskers plot, or box plot, have been attributed to John Tukey (1977).

```
1 |
2 | 7 7 8 8 8 9
3 | 1 1 2 2 2 2 3 4
4 | 1 1 3 4 5 5 6
5 | 0 1 2 2
6 |
```

Stem Plot

Transparencies 21.g and 21.h

Box Plots

Box plots use five summary numbers (lower extreme, lower quartile, median, upper quartile, and upper extreme) and are helpful when analyzing large quantities of data (more than 100 pieces of data). Although this type of display may be more difficult to construct, it has been used effectively with middle-school students.

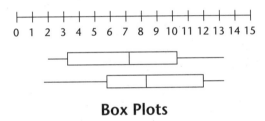

Box Plots

Transparencies 21.l and 21.q

Histograms

Histograms are block-graph representations of frequency tables of continuous data. The height of each bar is the frequency (count) of the variable being graphed. Because the graph deals with continuous data, the columns must touch; class intervals are used to designate width of the columns. The horizontal axis functions as a number line for noting the intervals.

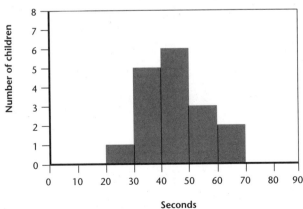

Transparency 21.m

Scatter Plots

Scatter plots are made using a two-dimensional array of points (quadrant A in a coordinate graph). Two variables are involved, one shown on each of the two axes. When possible, assess which of the variables is likely to be independent of the other. The independent variable is shown on the horizontal axis; the dependent variable (the variable that is likely to depend on the other) is shown on the vertical axis.

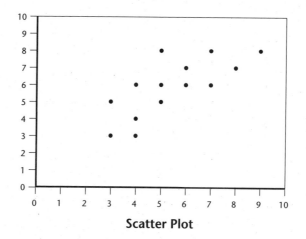

Scatter Plot

Distinguishing Among Bar Charts, Bar Graphs, and Histograms

Transparencies 21.a, 21.b, and 21.q

Three different kinds of graphs make use of bars.

A *bar chart* is used to show ungrouped data; a bar exists for each data item being represented. Spreadsheet software can often be used to create bar charts for entered data.

A *bar graph* is used to show grouped data; the data are counted in some "discrete" way so bars that do not touch can be displayed, showing frequencies for each data item. To show a bar graph with spreadsheet software, a frequency table must be entered; the software will not automatically group like items in a list of data items.

A *histogram* is used to show grouped data that are continuous or are so spread out that patterns will not emerge in a bar graph. The bars generally touch; they represent intervals in which the data items fall. It is difficult to create a histogram with standard spreadsheet software.

Deciding Which Graphs to Use

When deciding which graphs to use to represent data, we generally use two criteria. First, we consider what sort of data we have—the nature of the data: univariate or bivariate, categorical or numerical (discrete or continuous). Second, we look at why the data have been collected—the purposes of data collection: to describe, compare, or generalize (to make predictions about the next event or to draw conclusions about a population).

If the data are univariate, all but the scatter plot may be used; the scatter plot is used to represent bivariate data. The line graph and the histogram are used to display continuous data; the box plot and stem plot are used for both continuous and discrete data. Other graphs are used with discrete data.

With the exception of the box plot, all graphs may be used to describe data. All graphs, including the box plot, may be used to compare two or more data sets; this assumes there are two or more of a given representation so that there can be comparisons between or among samples. All graphs may be used as evidence for drawing conclusions or making predictions.

Levels of Graph Comprehension

While it is important for students to be able to read data presented in graphical form, a greater potential is realized when they are able to interpret and generalize from the data. Regardless of the graph form, there are three levels of graph comprehension.

Level 1: Reading the data. This level of comprehension requires a literal reading of the graph. The reader simply "lifts" facts explicitly stated in the graph, or information from the graph title and axis labels. There is no interpretation at this level; this type of comprehension is a low-level cognitive task.

Level 2: Reading between the data. This level of comprehension includes the interpretation and integration of data in the graph. It requires the ability to compare quantities (such as greater than, tallest, smallest) and the use of other mathematical concepts and skills (for example, addition, subtraction, multiplication, division) that allow the reader to combine and integrate data and to identify the mathematical relationships expressed in the graph. This is the level of comprehension most often assessed on standardized tests.

Level 3: Reading beyond the data. This level of comprehension requires the reader to predict or infer from the data by tapping existing schemata (such as background knowledge and memory) for information that is neither explicitly nor implicitly stated in the graph. Whereas reading between the data might require that the reader make an inference based on the data in the graph, reading beyond the data requires that the inference be made on the basis of a "database" in the reader's head.

As students experience the physical creation of a graph, they should be involved in interpreting it. Questions that reflect reading the data, reading between the data, and reading beyond the data provide a basis for interpreting and discussing graphs.

Discussions about graphs should revolve around listening, speaking, reading, and writing. Activities that provide students with the opportunity to interpret graphs and plots should include teacher-made and student-formulated questions reflecting different levels of comprehension. Students should be encouraged to write about graphs to clarify their thinking and to communicate their interpretations with others. Working in groups of four or five, they should talk about the graphs they create.

Using Raw Data or Reduced Data

When students are involved in a statistical investigation, an initial step in the analysis phase is making one or more representations of the data. This involves decisions about whether the data will be left ungrouped representing individual elements (raw data) or grouped in some way (reduced data). Data reduction can be achieved either by grouping the data values (such as by tallying the times each data value occurs) or by grouping the data values by the characteristic(s) of some attribute (such as by tallying the responses as they relate to gender). There are a variety of ways to represent data, and students may experiment with several in the course of examining a data set. Most choices must eventually focus on data reduction: Do we want to use a line plot? a bar graph? a stem plot? a histogram?

The decision about grouping has implications for the way a representation of data is made and interpreted. When data are ungrouped, each data value has its own plot element (for example, a bar).

Transparencies 21.a and 21.b

William says, "On one graph my bar is the tallest [21.a], but on the other graph my bar is one of the smallest [21.b]. Why?" When the data come from students in the class, each student can identify his or her bar. The "sizing" of each bar is not constant across data values. For example, a value of 10 is represented with a bar that is twice as tall as a bar showing a value of 5.

When data are grouped, the individual data values are "buried" within plot elements. For example, in a bar graph, all of the values of 10 are represented collectively in a single bar. The size of the bar is related to the number of 10s in the data set. If the number of 5s and the number of 10s are the same, the bars for showing the frequencies of these data values are the same height, though the data values themselves are not the same. Once we begin to reduce data, each data value is associated with a common-size unit (for example, a mark on a line plot or a bar on a bar graph). The unit of measure for sizing the plot elements is constant across data values.

The role of the axes in bar graphs depends on whether the data are ungrouped or grouped. For raw data, the horizontal axis is used for labeling the data values that name the bars. The vertical axis provides the actual data values when matched with the heights of the bars. When data are reduced with simple tallying, the horizontal axis provides the actual data values. Each of these values may have a bar associated with it. The vertical axis provides the frequency count of the occurrence of each of these values when matched with the height of each bar. Observe that the roles of the vertical axis for raw data and the horizontal axis for reduced data are the same. If students are to relate representations showing raw and reduced data, they must be aware of the change of perspective required as representations change.

Tables may play an important role as an intervening representation that can smooth the transition between representing raw and reduced data. Tallying may help make the distinction between ungrouped and grouped data clearer, since creating tallies of data values helps students focus on the relationships between data values and their frequencies. Tallying may also be useful in developing a sense of reversibility in data representations. Although tallies are usually made from raw data, students may also be asked whether they can "reconstruct" the original data set from the graphical display(s). With line plots and bar graphs, this is possible; with histograms and box plots, it is not. This may help students in their awareness and understanding of the subtleties that distinguish one representation from another.

The transition from ungrouped or raw data displayed on bar graphs to grouped data displayed on bar graphs, stem plots, and histograms seems to be appropriate for helping students to understand strategies for reducing increasingly complex data. For example, family size data (with a usual range of 10 people) are best represented in a standard line plot or bar graph. Breath-holding data (with a usual range of 100 seconds) are better represented in a stem plot than a bar graph. Data about time spent watching television in an evening (one class's range was over 480 minutes) are better represented using a histogram than a stem plot. Each of these representations responds to the need to identify patterns, even though the data may have increasingly greater spread or variability. Students need assistance to understand the types of data for which these representations are most applicable.

Summary of the Transparencies

The transparencies contain graphs and data that can be used to discuss the different kinds of graphs. Below are descriptions of each transparency (it is not intended that you use them in the order in which they appear). After the descriptions, is a summary by graph type.

Trans-parency	Description
21.a	Bar chart of ungrouped data about family size for 12 students
21.b	Bar graph of grouped data about family size for 12 students, displayed both with horizontal bars and vertical bars. You might have participants compare axis labels on the bar chart (21.a) and the bar graphs (21.b).
21.c	Bar graphs of summary data from a survey question displaying categorical data; response frequencies are displayed as percentages so that data from two displays can be compared without reference to group size.
21.d	Line graph showing height over time of S. Friel's niece. Notice that age is on the horizontal axis. Can participants determine how tall Meg was at age 2 (interpolation)? How tall she will she be at age 10 (extrapolation)?
21.e	Double line graph showing average heights of males and females over time. You can ask interpolation and extrapolation questions here as well.
21.f	Data sheet for hot dog graphs. You may want to distribute a copy to participants.
21.g	Box plot of calories in different kinds of hot dogs.
21.h	Box plot of sodium content in different kinds of hot dogs. Notice the outlier.
21.i	Data sheet for travel time and distance to school for a group of students.
21.j	Bar graph of categorical data on mode of travel, and circle graph of categorical data on mode of travel with percentages shown.
21.k	Travel time to school as ungrouped data in a bar chart and as grouped data in bar graph by simple count of frequencies. Have participants note the problems with scale on the horizontal axis.

21.l Travel time to school as grouped data in a bar graph that shows tallies of frequencies within given intervals. This is a step before the histogram and is not often used. Some computer software does it this way, but the histogram shown is "right" for our purposes.

21.m Scatter plot showing relationship of travel time to distance.

21.n Data sheet for travel time and distance to school for a second group of students.

21.o Beginning of stem plots for travel time to school for the second group of students, both unordered and ordered. You may want participants to complete the stem plots and also construct a stem plot for distance traveled.

21.p Data sheet of animal pulse rates (full data set is available in *Used Numbers: Middles, Means, and In-Betweens*).

21.q Three different histograms of the animal pulse data, permitting comparisons of different interval options.

Summary of Graph Types	
Graph type	Transparency
Pictograph	none
Bar chart	21.a, 21.k
Bar graph	21.b, 21.c, 21.j, 21.k, 21.l
Circle graph	21.j
Line graph	21.d, 21.e
Line plot	none
Stem plot	21.o
Histogram	21.l, 21.q
Scatter plot	21.m
Box plot	21.g, 21.h

References

Curcio, Francis R. *Developing Graph Comprehension*. Reston, Virginia: National Council of Teachers of Mathematics, 1989.

Friel, S. N., J. R. Mokros, and S. J. Russell, *Statistics: Middles, Means, and In-Betweens*. Palo Alto, California: Dale Seymour Publications, 1992.

Family Size Bar Chart: Ungrouped Data (Unordered)

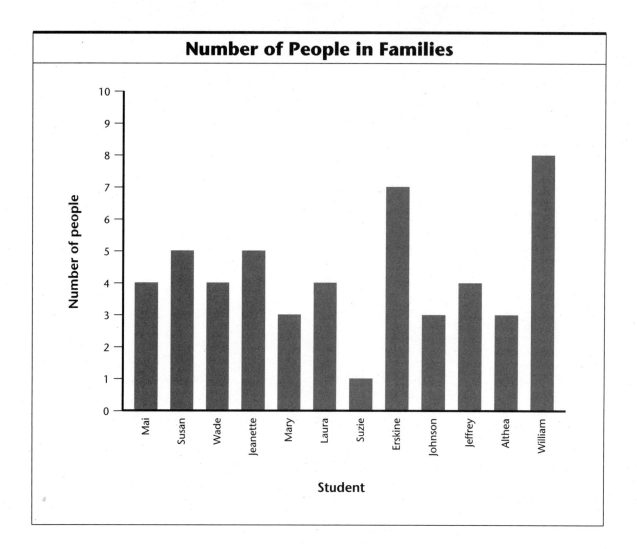

Family Size Bar Graphs: Grouped Data

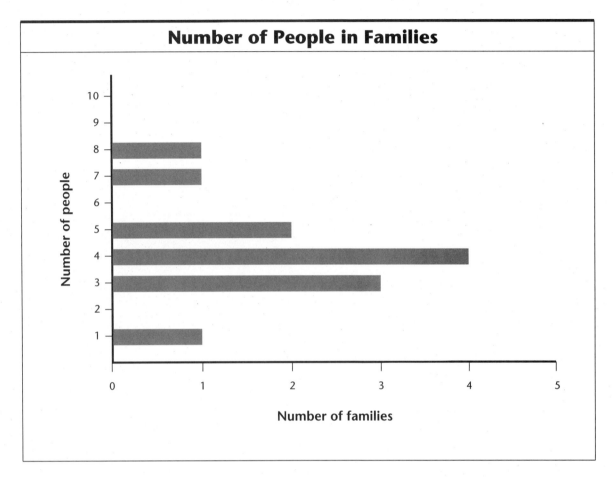

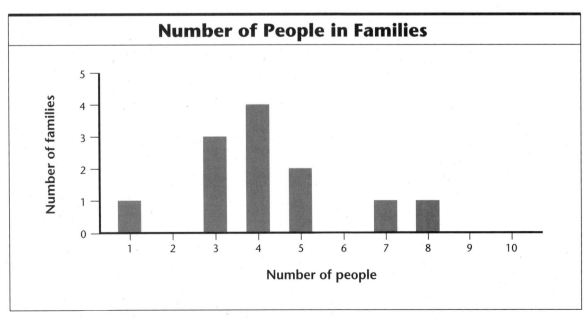

Student Activities Bar Graphs

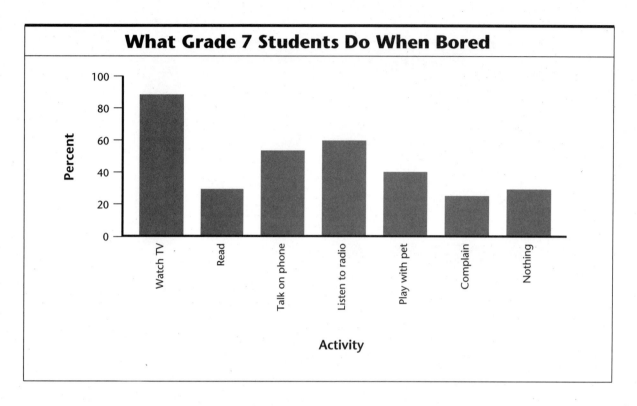

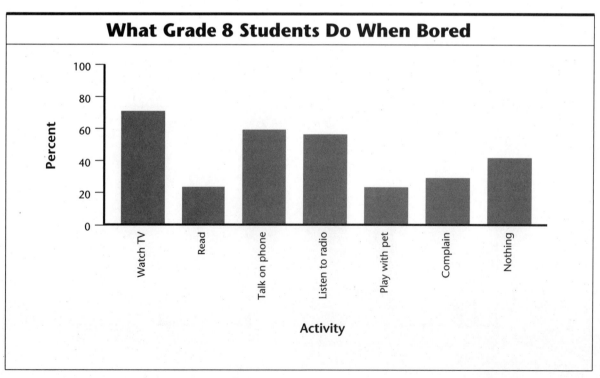

Child's Growth Line Plot

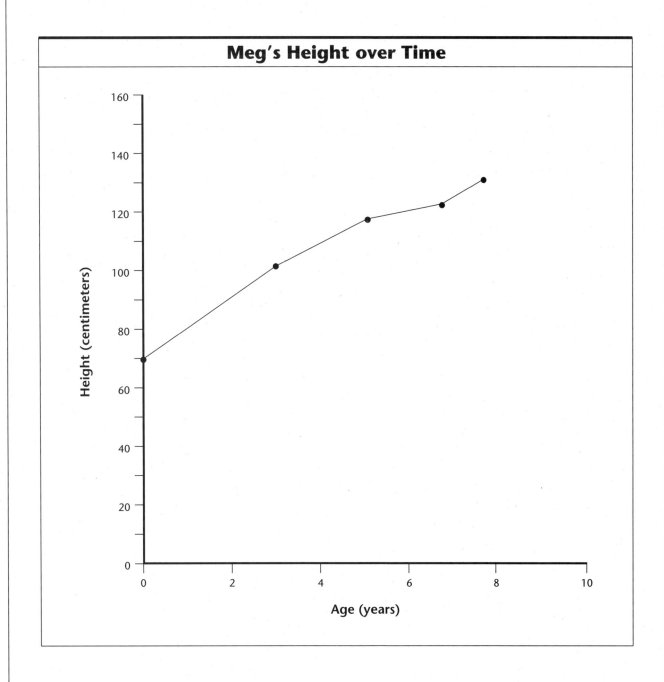

Average Height Double Line Graph

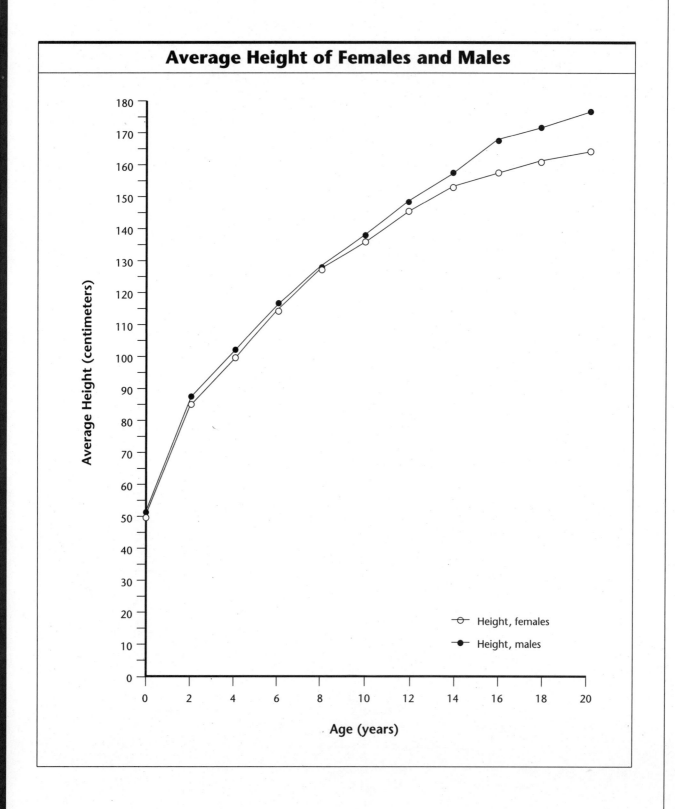

Hot Dogs Database

	Hot dog type	Calories	Sodium
1	beef	186	495
2	beef	181	477
3	beef	176	425
4	beef	149	322
5	beef	184	482
6	beef	190	587
7	beef	158	370
8	beef	139	322
9	beef	175	479
10	beef	148	375
11	beef	152	330
12	beef	111	300
13	beef	141	386
14	beef	153	401
15	beef	190	645
16	beef	157	440
17	beef	131	317
18	beef	149	319
19	beef	135	298
20	beef	132	253
21	meat	173	458
22	meat	191	506
23	meat	182	473
24	meat	190	545
25	meat	172	496
26	meat	147	360
27	meat	146	387
28	meat	139	386
29	meat	175	507
30	meat	136	393
31	meat	179	405
32	meat	153	372
33	meat	107	144
34	meat	195	511
35	meat	135	405
36	meat	140	428
37	meat	138	339
38	poultry	129	430
39	poultry	132	375
40	poultry	102	396
41	poultry	106	383
42	poultry	94	387
43	poultry	102	542
44	poultry	87	359
45	poultry	99	357
46	poultry	170	528
47	poultry	113	513
48	poultry	135	426
49	poultry	142	513
50	poultry	86	358
51	poultry	143	581
52	poultry	152	588
53	poultry	146	522
54	poultry	144	545

Hot Dog Calories Box Plots

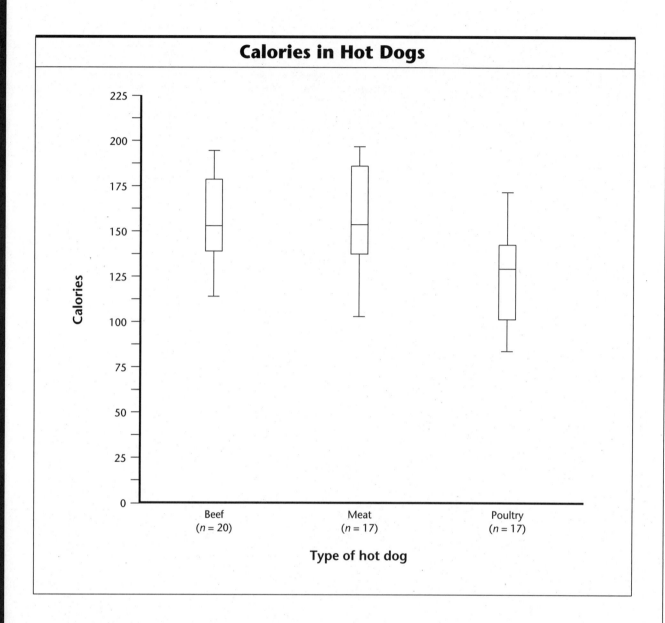

Hot Dog Sodium Content Box Plot

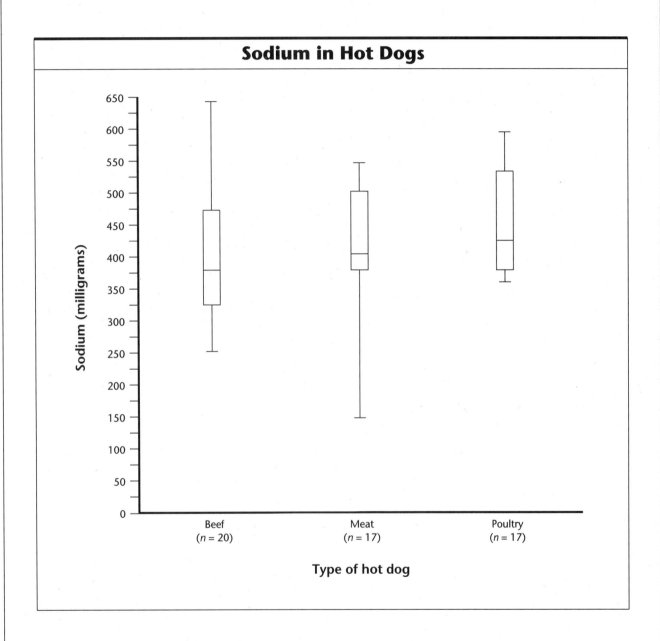

Travel Time, Distance, and Mode Database 1

Student	Time (minutes)	Distance (miles)	Mode of travel
LS	29	1.5	Bus
DS	15	1.25	Bus
RV	25	1.25	Bus
DB	15	1.25	Bus
HR	30	2.5	Bus
MD	10	1.25	Bus
SY	30	3	Bus
QM	35	2	Bus
KB	15	1.25	Bus
JC	25	1.25	Bus
ML	22	2.5	Bus
JG	13	1.5	Bus
CF	12	1	Car
CB	23	1.25	Bus
NA	20	1	Bus
CS	27	2.5	Bus
DC	23	1.25	Bus
JS	30	1.25	Bus
AP	25	1.5	Bus
GA	20	1	Bus
DG	28	0.75	Bus
MT	23	1.25	Bus
EA	30	3	Bus
EF	21	1	Bus
JR	29	1.5	Bus
MF	12	1	Car
KA	15	1.25	Foot
CA	18	0.75	Foot
AM	15	0.5	Foot
KW	18	1	Foot
ID	3	0.25	Foot
WW	9	0.5	Foot
MM	24	1.5	Foot
TS	17	1	Foot
TR	45	2	Foot
AR	5	0.25	Foot
FL	7	0.25	Foot
CG	18	1	Foot
MG	8	0.25	Foot
AM	16	0.5	Foot
CJ	19	1.25	Foot

Mode of Travel Bar Graph and Circle Graph

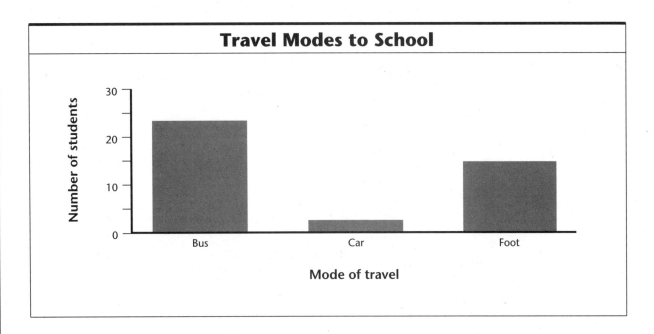

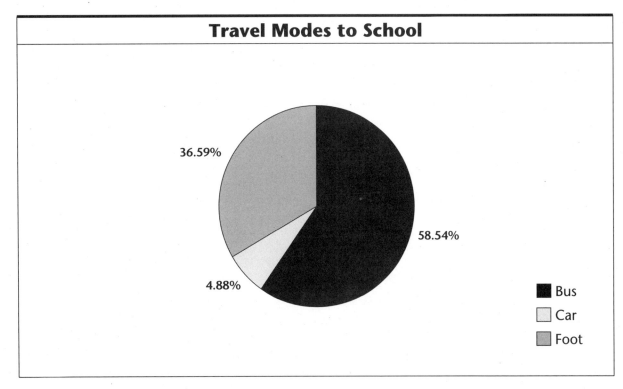

Travel Time Bar Chart: Grouped and Ungrouped Data

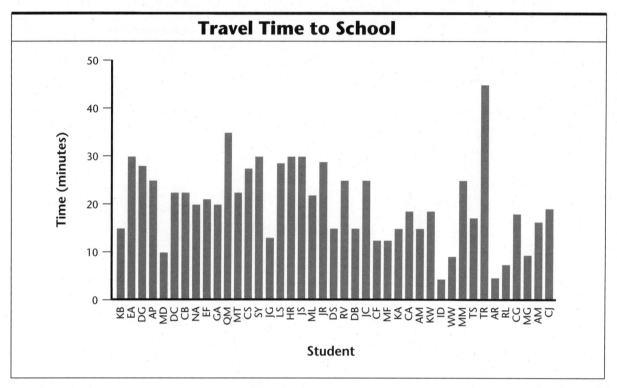

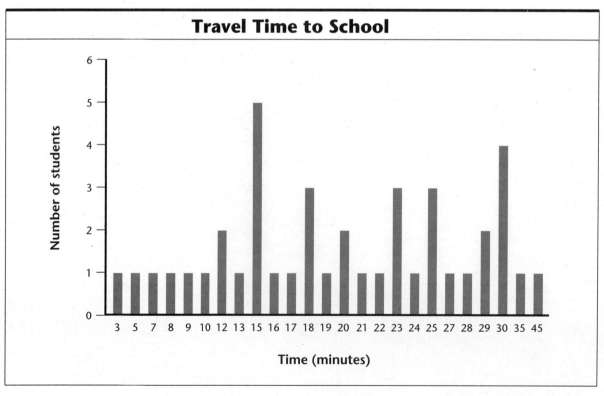

Travel Time Bar Graph and Histogram: Grouped Data

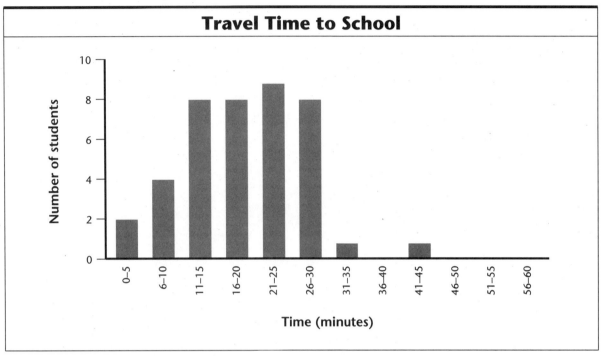

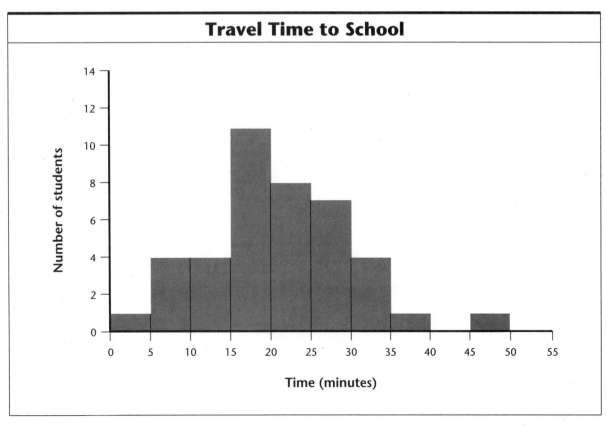

Travel Time and Distance Scatter Plot

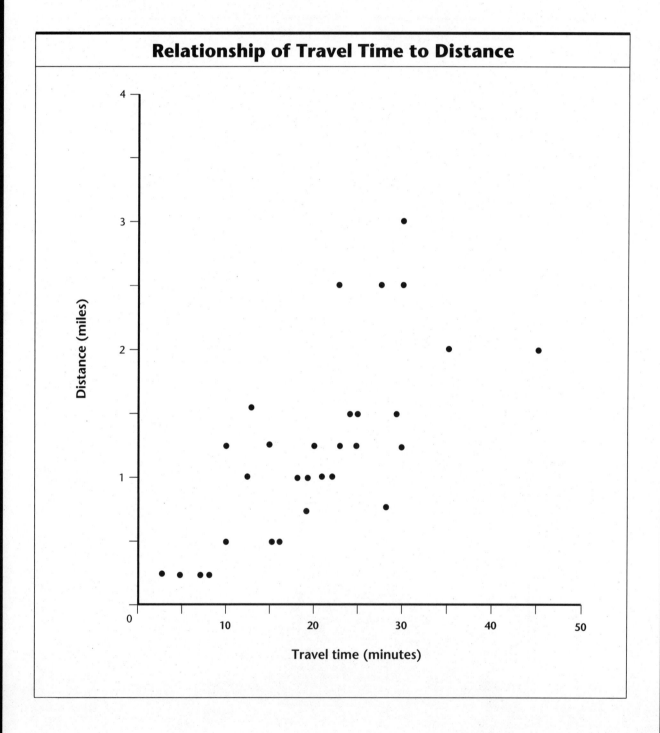

Travel Time, Distance, and Mode Database 2

Student	Time (minutes)	Distance (miles)	Mode of travel
DB	60	4.7	Bus
CC	30	1.9	Bus
DD	15	1.9	Bus
SE	15	0.8	Car
AE	15	1.0	Bus
FH	35	2.6	Bus
CL	15	1.1	Bus
LM	22	2.0	Bus
QN	25	1.6	Bus
MP	20	1.5	Bus
AP	25	1.3	Bus
AP	19	2.2	Bus
HCP	15	1.4	Bus
KR	8	0.2	Foot
NS	8	1.2	Car
LS	5	0.6	Bus
AT	20	2.8	Bus
JW	15	1.6	Bus
DW	17	2.4	Bus
SW	15	1.9	Car
NW	10	0.5	Foot
JW	20	0.6	Foot
CW	15	2.2	Bus
BA	30	3.0	Bus
JB	20	2.6	Bus
AB	50	4.0	Bus
BB	30	4.8	Bus
MB	20	2.0	Bus
RC	10	1.3	Bus
CD	5	0.3	Foot
ME	5	0.4	Bus
CF	20	1.7	Bus
KG	15	1.8	Bus
TH	11	1.6	Bus
EL	6	1.2	Car
KLD	35	0.8	Bus
MN	17	4.5	Bus
JO	10	3.1	Car
RP	21	1.6	Bus
ER	10	1.0	Bus

Travel Time Stem Plots

Initial Version		Reordered Version	
0	8 8 5	0	5 8 8
1	5 5 5 5 9 5 5 7 5	1	5 5 5 5 5 5 5 7 9
2	2 5 0 5 0	2	0 0 2 5 5
3	0 5	3	0 5
4		4	
5		5	
6	0	6	0
7		7	
8		8	
9		9	

2|5 means 25 minutes

Animal Pulse Rates Database

Animal	Pulse rate (beats per minute)	Animal	Pulse rate (beats per minute)
Alligator	47	Groundhog	80
Bass	128	Guinea pig	270
Basset hound	92	Haddock	37
Beagle	120	Horse	35
Bear	55	Human (adult)	70
Beaver	140	Human (baby)	148
Boston terrier	120	Irish terrier	100
Camel	25	Kangaroo	120
Carp	59	Lion	42
Cat	130	Mule	40
Cod	48	Ostrich	70
Collie	100	Pekingese	125
Cow	55	Perch	59
Crocodile	48	Pig	70
Deer	70	Pigeon	170
Dolphin	110	Pointer	90
Donkey	40	Porcupine	300
Duck	268	Rabbit	150
Elephant	30	Salmon	38
Fox	240	Seal	44
Fox terrier	95	Shark	29
Frog	48	Sheep	75
Giraffe	66	Trout	35
Goat	80	Turkey	211
Goldfish	38	Whale	16

Friel, S. N., J. R. Mokros, and S. J. Russell, *Statistics: Middles, Means, and In-Betweens*. Palo Alto, California: Dale Seymour Publications, 1992.

Animal Pulse Rates Histograms

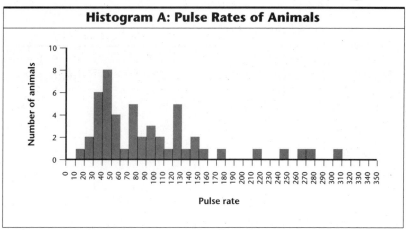

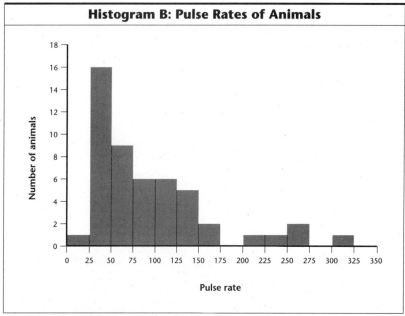

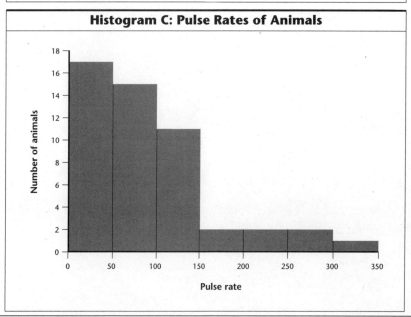

INVESTIGATION 22
Raisins Revisited

Overview

During Investigation 6, *Shape of the Data: Using Line Plots,* participants gathered data on one brand of raisins. This activity extends the raisin data set by the examination of other brands. Participants use the earlier information to make predictions about the new brands, then gather and represent data to evaluate their predictions. The new data are used to extend predictions to different-size boxes of the various brands of raisins. Finally, the data are represented using technology (computers and software, such as *Data Insights*, *MacS*tat, or *Data Wonder*). If there is only one computer available per room, this part of the activity may be conducted as a demonstration.

Prediction, which reappears in activities later in the workshop, depends heavily on a holistic understanding of what a representation communicates.

Assumptions

Participants are familiar with some types of representations (such as bar graph and tallies) but may need instruction on other types (such as histograms and box plots). Virtually all of the technology section may be new material for many participants.

Goals

Participants will describe ways to compare data. In particular, they

- explore the concept of sampling

- understand the appropriateness of various representations to display and compare samples

- describe the shape of data from different samples to make comparisons

- use measures of center and measures of spread to describe and compare samples

- realize that different interpretations of data are possible

Materials

15–20 half-ounce boxes of raisins of a single brand for each group of 3–6 participants (various brands represented), several boxes of raisins of other sizes, chart paper, flip chart (optional), balance scales (optional), computer(s) and statistical software (optional), calculators

Handout 22.1

Teacher Notes

If this activity seems too long, you may want to consider doing it partly as a demonstration or as a whole group and partly in small groups.

Fun Facts

Grapes are harvested for raisins when their sugar content is 19 percent or more. They are dried in the sun for 2 to 3 weeks until their moisture content is 14 or 15 percent, at which point they are wrapped for more drying.

Producing a single pound of raisins requires around 4½ pounds of grapes.

Developing the Activity: Part 1

Participants gather and analyze data on several brands of raisins, then make predictions for boxes of different sizes.

Pose the Question

The problem is posed to each group.

Earlier, you gathered data on one brand of raisins. Do you think all brands have similar numbers of raisins in their boxes? If not, what might account for differences in numbers of raisins across brands?

Brainstorm these ideas, recording responses on the board or overhead projector.

About how many raisins do you expect to find in boxes of each of these brands [list the brands]*? Why do you think so?*

Write the predictions for each brand. Ask participants to justify their predictions (for example, generic brands may be of lower quality and thus perhaps less moist, so more would fit in a box).

What similarities and differences do you notice in your predictions?

Let participants discuss any patterns they see. Since they have not made any representations of these data, it may be difficult for them to compare across brands. Ask how the comparisons might be made easier if representations were made of the data (however, don't spend time now making representations).

Collect the Data

Let each small group (three to six people) gather the data for one brand of raisins. Ask them to record the number of raisins on the lid of each box. (The boxes can later be stacked to make a "real" graph.) If a group finishes early, ask them to discuss the question of how many boxes would be needed to get a good sense of the raisin count for a particular brand.

As an alternative procedure, let participants open all boxes for one brand. Record the data on the board or flip chart. Then open boxes for the second brand and record the data, and so on.

Participants may want to weigh the raisins, either now or during the analysis phase. If balance scales are available, and if there is time, let them gather this additional data. Then work with those who are

212 *Raisins Revisited*

interested in using the additional data in the analysis (for example, to make scatter plots).

Analyze the Data

Ask each team to share its data, and write the data on chart paper taped to the wall so that everyone can see.

Ask how the data might be represented to compare the different brands, such as with bar charts, a multibar chart, a color-coded line plot, box plots, a real graph made from raisin boxes piled on a number line, and stem plots (or a multicolored stem plot). List the suggestions. Some participants won't know how to make all of these plots, so be prepared to demonstrate later how to make the representations that are least familiar. Don't force all these representations into the list during this discussion; just be prepared to demonstrate later.

Ask what information would be communicated by each of the following representations, and discuss the differences thoroughly:

- a bar chart, with one bar for each box

- a frequency distribution, with each bar representing the number of boxes for each count

- a histogram, with each bar representing the number of boxes in an interval (for example, 30–34 raisins)

Ask which representations would be easiest for comparisons, and make sure participants justify their choices. Be prepared to discuss and model the differences between a bar chart (ungrouped data) and a histogram (grouped data). Ask participants how the shape of the display changes when moving from a bar chart to a histogram.

Ask each group to sketch one of the representations, and have them display their work.

Interpret the Results

Discuss the information that can be read from the representations presented. Which brand has the tallest peak? The most clumping together? The greatest spread? Highlight some of the notions of variability—median, median of each half (interquartile ranges)—as these ideas will be needed during the technology part of the activity. Ask how the means of the various brands could be determined.

 Fun Facts

The history of raisins stretches back as far as 400 B.C., when people in the fertile lands of ancient Armenia cultivated grapes, which could easily be turned into raisins in their dry, warm region.

Fun Facts

Did you ever wonder why all the raisins don't end up at the bottom of the cereal box? The manufacturers' trick is to add the raisins only after more than half of the box has been filled; the raisins will be evenly distributed as they settle during shipping.

From Feldman, D. *Why Do Clocks Run Clockwise? and Other Imponderables.* New York: Harper and Row, 1987, p. 123.

What would this tell you about the brands?

If no one suggests using stem plots or box plots, present at least one as an alternative. Demonstrate its construction, and use it to compare data across brands.

Another representation of considerable interest can be made with the boxes themselves. Stack one brand of boxes in order according to the number of raisins per box to make a "real" bar graph. Stack a second brand on top of the first, and discuss the amount or lack of overlap of the two distributions. Continue building the bar graph with the remaining brands.

It is critical to discuss whether the sample of boxes for each brand is adequate to make comparisons. Possible concerns are where the raisins were bought, the time of year they were packaged, and the number of boxes examined. Discuss how—and how expensive it would be—to overcome the possible sampling bias, and whether we should or can draw conclusions when we don't have complete information (of course, in real life we do this all the time!).

Summary

Show participants different sizes of boxes of raisins and ask how they might predict the number of raisins in each box. Give two or three boxes to each small group, and let them make predictions. Discuss the predictions; some participants will probably have used proportional reasoning.

How important is it to know how accurate your predictions are?

Discuss this question, and ask participants to project how important it would be for their students. Be sure it is clear that each box of raisins is only one element in a sample; we cannot know how representative a single box is of the entire population of similar-size boxes. If participants express a need to verify their predictions, have them open and count the raisins in the larger boxes. Some participants will likely comment that the data from one box cannot validate—or invalidate—a prediction, because they can't know how representative one piece of data is for that brand.

Developing the Activity: Part 2

Participants will explore the use of technology—computers and software—to create representations of data. Precisely what happens here depends on the hardware and software that is available.

Demonstrate the use of technology for recording and representing data. Enter the data, and then demonstrate the representations that are available. If box plots are available, be sure to demonstrate and discuss them. Try to involve some of the participants in entering data and generating representations.

Discuss the advantages and disadvantages of using technology for data representation.

Summary

This problem will generate interesting discussion about how to represent data so that comparisons can be made conveniently. Questions about the representativeness of particular boxes of raisins are likely to be raised. Challenge participants to identify other objects that could easily be compared by students (for example, sheets of toilet paper per roll, kernels of popcorn in individual microwave bags, and chips in chocolate-chip cookies).

Fun Facts

The United States produces one third of the world's total raisin crop. Almost all of those raisins are produced in California's San Joaquin Valley.

HANDOUT 22.1

Raisins Problem Sheet

1. Do all brands of raisins have the same number of raisins per box? Explain your answer.

2. List the brand(s) available, and predict how many raisins you think will be in each box.

    Brand   **Prediction**

3. Write your guesses on the class chart. What patterns do you notice in the predictions?

4. As a team, gather data for one brand of raisins. Write your data on the class chart, then make several representations of your data.

5. What conclusions can you draw about your brand of raisins?

6. How many raisins do you think there would be in a 3-ounce box of raisins? in an 8-ounce box of raisins?

INVESTIGATION 23

How Do We Grow?

Overview

In this activity, participants explore what it means to compare data sets using either the same kind of representation (for example, two bar graphs, one bar graph for each data set) or using box plots. The problem is couched in an exploration of how people grow. Participants consider data for groups of people of different ages to describe how people grow over time.

Assumptions

Participants are familiar with concepts related to describing the shape of the data and with different ways to represent data, including line plots, bar graphs, stem plots, histograms, and line graphs. They have an understanding of the primary purposes of data collection: to describe data, to compare data sets, and to generalize from a sample to the next case or to the population.

Goals

Participants consider issues related to describing the shape of data. In particular, they

- consider ways to compare data: using the same kind of representations; using knowledge about the shape of the data as a basis for discussion of similarities and differences

- learn how to make and use box plots as tools to compare data

- consider both interpolation and extrapolation of information using line graphs

References

Landwehr, J. M., and A. E. Watkins. *Exploring Data*. Rev. ed. Palo Alto, California: Dale Seymour Publications, 1995.

Russell, S. J., and R. B. Corwin. *Statistics: The Shape of the Data*. Palo Alto, California: Dale Seymour Publications, 1989.

Materials

Graph paper, calculators

Transparencies 23.a through 23.h

Handouts 23.1 through 23.3

Fey, J., W. Fitzgerald, S. Friel, G. Lappan, and E. Phillips. *Data Around Us*. Palo Alto, California: Dale Seymour Publications. Forthcoming.

Developing the Activity: Part 1

Participants compare two data sets, each represented on a bar graph.

Pose the Question

Using Transparency 23.a (both on the overhead projector and as a handout for all participants), request that participants compare the two sets of data.

The data represented in the bar graphs are about two grade 4 classes—one in Massachusetts and one in Georgia. We want to compare how these two classes are alike and different, looking for ways to characterize the students' height. We also want to raise any questions that surface as a result of our comparisons.

Collect and Analyze the Data

Have participants, working in small groups, discuss the two data sets. Some groups may immediately focus on the obvious discrepancies in height between the classes, making statements such as, "The Georgia class may have some students who are older." Help participants understand that they cannot determine this sort of information from the data as they are presented, and comment on the natural tendency to want to explain data. However, we simply want to describe observations about the data, using our descriptions to identify ways the two data sets are similar and different. One comparison, for example, emerges from the following description:

In the Massachusetts class the shortest students are 51 inches tall, and the tallest are 58 inches. In the Georgia class the shortest students are 51 inches tall, and the tallest are 64 inches. The shortest students in the Massachusetts class are as short as the shortest in the Georgia class, but the tallest Georgia students are taller than the tallest Massachusetts students.

Participants may want to talk about the peaks in the distributions, the number of peaks, and the medians.

Interpret the Results

Once you have compared the data for the two groups based on what can be observed, discuss some of the hypotheses participants have for why the shapes of the height data for the classes are so different. Participants may identify other variables that could be investigated (in other words, pose additional questions).

At some point, let participants know that the data for the Massachusetts class were collected in the fall of the school year and that of the Georgia class were collected in the spring. This explains much of what is happening in the data, and also introduces the notion of growth. You may want to have participants discuss the following question, raising their own questions as they talk.

What do we know about patterns of growth over time, and how might we explore patterns of growth?

Participants may raise many issues. Can they describe times of fast growth and times of slow growth? When are people finished growing? How tall do people typically get?

Summary

Comment on why comparisons can be made between the two bar graphs. Note that the two data sets are not the same size; this is often of concern to students. What you are comparing are the *characteristics* of the shape of the data: center, spread, peaks, symmetry or skewness, clusters, outliers. Through a discussion of the shape of the data, ways for comparing the data should emerge that do not necessitate having the same number of data items.

In this investigation, the problem is not one of looking for ways to characterize grade 4 students in terms of their height, but one that involves looking for ways to characterize *growth* of grade 4 students in terms of their height. Of course, the way the problem was posed—deliberately—encouraged participants to address the former focus, not the latter.

Help participants realize that the scale of the vertical axes on the bar graphs must be the same to be able to make comparisons across the two representations. Display Transparencies 23.b and 23.c, which show three graphs of raisin data that have problems with scale on both the horizontal and vertical axes (23.b) and then have the problems corrected (23.c). The corrected graphs take into account sample size by showing data as a percent of the total sample (or relative frequency). This makes comparisons across the three samples possible. However, note that the A&P raisin sample is quite small; this may not be enough data to make a fair comparison.

Developing the Activity: Part 2

Participants learn to construct five-number summaries and box plots, and they practice interpreting box plots.

Pose the Question

We want to look at a different way to compare the two sets of data about student height. First, we need to identify what is known as the "five-number summary" for each data set: the median, the upper and lower quartiles, and the minimum and maximum values.

Collect and Analyze the Data

Finding Five-Number Summaries

Display Transparency 23.d, which shows only the Massachusetts data.

Here is the bar graph for the Massachusetts class and a list of the individual heights in order from smallest to largest. To create our five-number summary, we first identify the median.

Participants should be able to do this easily. There are an even number of heights (18), so the median is located between the ninth and tenth heights (halfway between 54 and 55, or at 54.5).

Introduce participants to two other "midpoints": the *lower quartile* (the median of the numbers at or below the median) and the *upper quartile* (the median of the numbers at or above the median). The lower quartile is found for the heights of 51–54 inches; it is the fifth value in the ordered sequence of 18 heights, or 53 inches. The upper quartile is found for the heights of 55–58 inches; it is the fourteenth value in the ordered sequence of 18 heights, or 56 inches.

Finally, participants can identify and locate the *minimum* (lowest) and *maximum* (highest) values, completing the five-number summary for the Massachusetts data.

Minimum	51
Lower quartile	53
Median	54.5
Upper quartile	56
Maximum	58

Display Transparency 23.e—the bar graph and ordered list of heights for the Georgia class—and have participants identify the five-number summary for this data. Because this data set has an odd number of values (21), the median is the middle height, or 57 inches. In determining the quartiles, the height that is the actual median is not counted as part of the lower or upper halves of the data.

Minimum	51
Lower quartile	54
Median	57
Upper quartile	59.5
Maximum	64

Making Box Plots Using Five-Number Summaries

Returning to the Massachusetts data, we can demonstrate how to make a box-and-whiskers plot, or box plot, using the five-number summary. See the completed summary in Transparency 23.f. The box plot can be made below the horizontal axis of the bar graph by locating the median and then the two quartiles and drawing the "box" of the plot. The "whiskers" are drawn to the minimum and maximum values.

What do we know about the data related to the box plot?

Fifty percent of the data is contained in the box, and 25 percent of the data is accounted for by each of the two whiskers.

Have participants create the box plot for the Georgia data. You can show Transparency 23.g to summarize their work.

Checking for Outliers

There will sometimes be unusually high or low values in a data set, which are called *outliers*. When constructing box plots, there are strategies for identifying these values.

Statisticians have different ways to identify outliers. We will use the "one and a half" rule. We take 1.5 times the length of the box in the box plot, which is known as the **interquartile range.** *For the Massachusetts data, the box length is 3 inches (53 to 56). Multiplying this by 1.5 gives us 4.5 inches. This means that the whiskers on each side of the box may not be drawn any longer than a distance of 6 inches from the box. If a value falls beyond this distance from the box, it is considered an outlier and is marked with an asterisk. Do we have any outliers in the Massachusetts data?*

There are no outliers in the Massachusetts data. Give participants time to determine the cutoff for the Georgia data (5 × 1.5 = 7.5 in.) and to identify whether there are any outliers (there are not).

Teacher Notes

Statisticians differ on their opinions about whether the median should be included as a member of each of the two halves of the data when determining quartiles. We will not include it in our computations.

Interpret the Results

We can use box plots to display our data and to make comparisons.

Using Transparency 23.h, discuss how box plots help us to see the likenesses and differences between two data sets.

Using these two box plots, in what ways can we describe how students in grade 4 grow? What can we say about where 50 percent of the data fall (that is, the two boxes—both the two sets of quartiles and the interquartile range for each)? What about the minimum and maximum heights? The median heights?

Summary

The box plot is somewhat complicated to construct, since you must first calculate the median, extremes, and quartiles. An advantage of the box plot is that it highlights only a few important features of the data, focusing attention on the median, extremes, and quartiles, and comparisons among them. Another advantage is that the box plot is useful with any number of values. A disadvantage occurs when there are only a few data values—fewer than about 15. In such a small sample, the plotted values might change greatly if only one or a few of the observations were altered or additional data were included.

Developing the Activity: Part 3

Participants consider other data sets about people's height, add their own data to one of the sets, construct five-number summaries and box plots of the data sets, and make comparisons.

Pose the Question

Distribute Handout 23.1.

Here we have a number of data sets of information about the height of different age groups. Add your data, in centimeters, to the group "Adults." We want to make a general display of box plots for each of the data sets. Then, using this information, we will discuss what we know about how people grow.

Collect and Analyze the Data

Have participants determine the five-number summaries and make box plots for the data on the handout. Show all the box plots on the same display. If participants want, they can convert the grade 4 data from inches to centimeters (using the conversion factor of 2.5 and rounding consistently where needed) and add them to their display.

Interpret the Results

Have participants work in small groups and use the displays of data to respond to this question:

How do people grow over time?

Participants should record all observations and then add further questions or requirements for information they feel would be necessary to redo their analysis with more accuracy and thoroughness.

Height data for three different groups of basketball players are provided in Handout 23.2. Participants may want to add that data to their display and discuss two questions: What is typical about basketball players' heights? What does the addition of these heights to our height data tell us about human growth? They may want to consider other questions raised during their consideration of the data.

Summary

Point out to participants the differences in the sizes of the data sets and how the use of box plots makes comparisons possible. Comparing raw data—even considering the shapes of the distributions—would be difficult using this many data sets and involve a wide range in the numbers of people in a given data set.

Participants may want to add their students' data or the data from other classrooms in their school to this data set. They will need to collect data from all groups at the same time (recall the problem with the two grade 4 classes).

As another facet of this problem, you may want to ask participants when they think people attain their adult height.

Massachusetts and Georgia Classes Bar Graphs

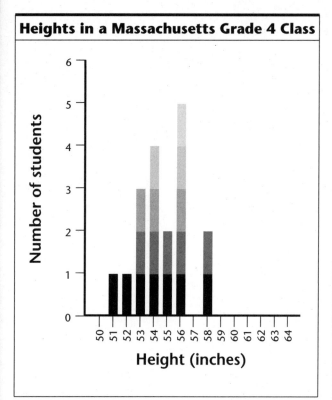

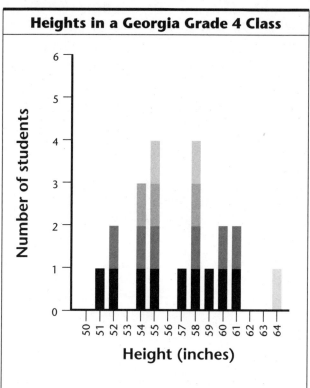

Russell, S. J., and R. B. Corwin. *Statistics: The Shape of the Data.* Palo Alto, California: Dale Seymour Publications, 1989.

Raisins Histograms (Uncorrected Scales)

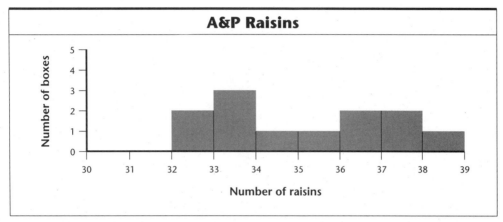

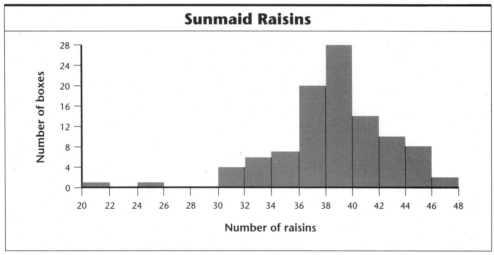

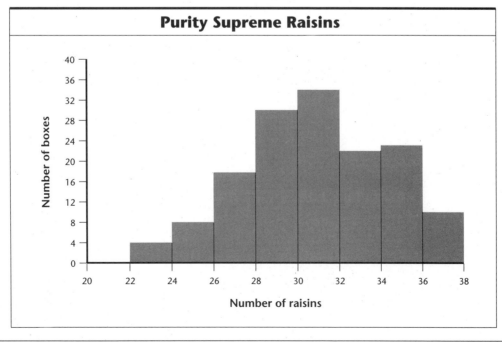

Raisins Histograms
(Corrected Scales)

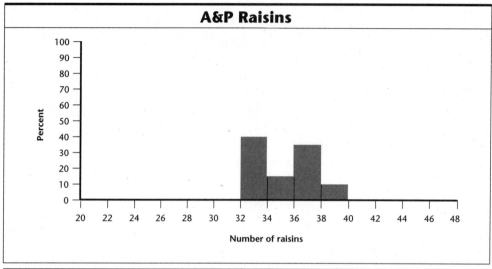

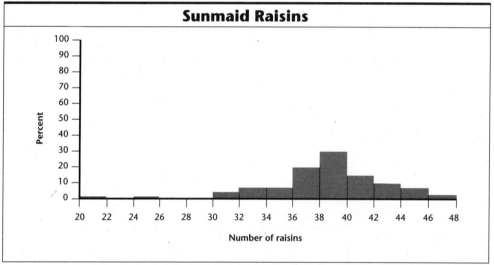

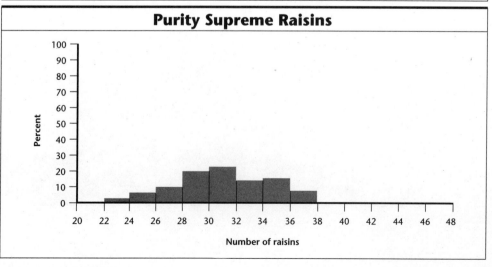

Massachusetts Class Bar Graph and Data Values

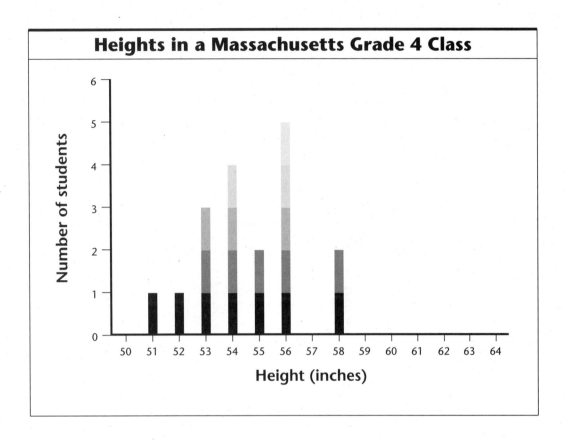

Students' Heights

51 52 53 53 53 54 54 54 54
55 55 56 56 56 56 56 58 58

Five-Number Summary

Minimum:

Lower quartile:

Median:

Upper quartile:

Maximum:

Georgia Class Bar Graph and Data Values

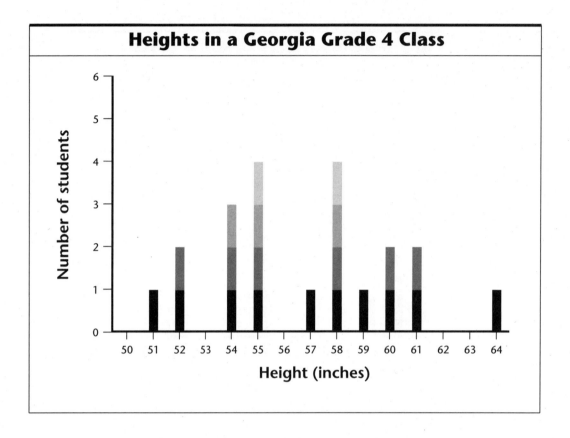

Students' Heights

51 52 52 54 54 54 55 55 55 55 57
58 58 58 58 59 60 60 61 61 64

Five-Number Summary

Minimum:

Lower quartile:

Median:

Upper quartile:

Maximum:

Massachusetts Class Bar Graph, Data Values, and Box Plot

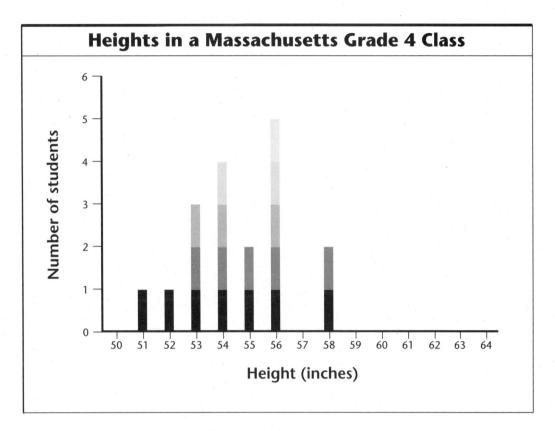

Box Plot

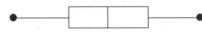

Students' Heights

51 52 53 53 53 54 54 54 54
55 55 56 56 56 56 56 58 58

Five-Number Summary

Minimum: 51
Lower quartile: 53
Median: 54.5
Upper quartile: 56
Maximum: 58

Georgia Class Bar Graph, Data Values, and Box Plot

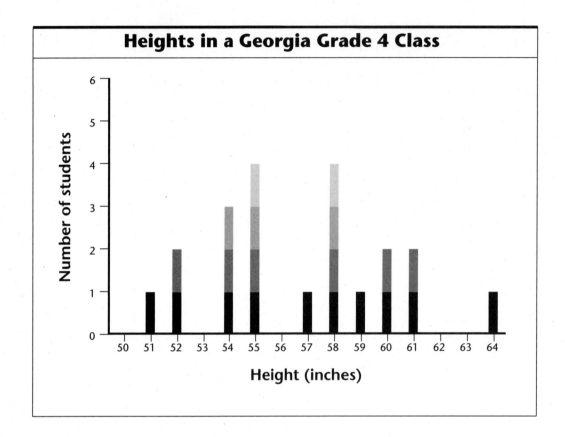

Box Plot

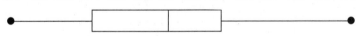

Students' Heights

51 52 52 54 54 54 55 55 55 55 57
58 58 58 58 59 60 60 61 61 64

Five-Number Summary

Minimum: 51
Lower quartile: 54
Median: 57
Upper quartile: 59.5
Maximum: 64

Massachusetts and Georgia Classes Box Plots

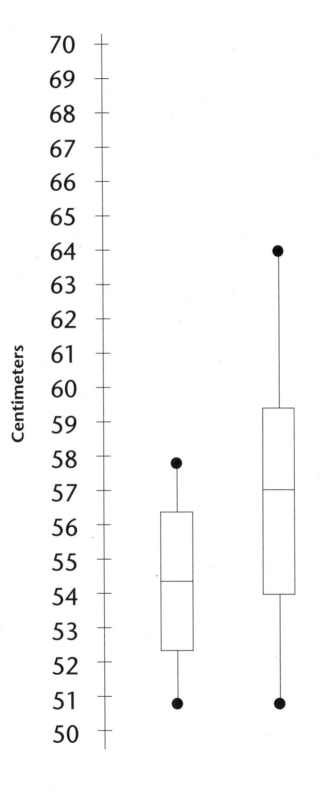

Massachusetts and Georgia Classes Bar Graphs

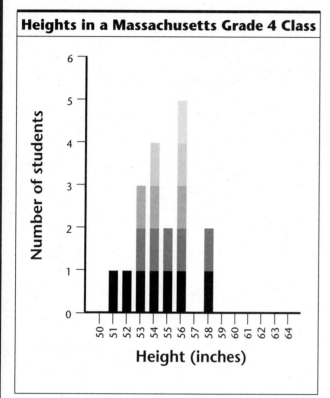

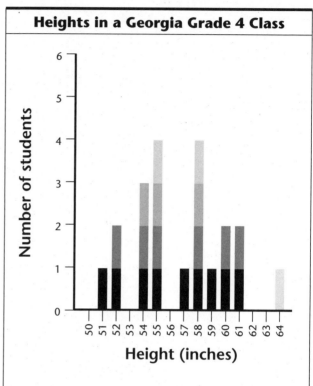

Russell, S. J., and R. B. Corwin. *Statistics: The Shape of the Data.* Palo Alto, California: Dale Seymour Publications, 1989.

Height Data, Sheet 1

		Heights of Different Groups (centimeters)			
Newborns	Grade 1 students	Grade 5 students	Grade 8 students	Adults	
43	117	138	147	145	168
45	118	138	156	148	168
45	119	138	159	150	168
46	120	139	160	150	170
49	122	141	160	153	170
49	123	142	161	153	170
50	125	144	162	155	170
50	125	146	162	155	170
50	127	147	162	155	172
51	127	147	163	158	172
53	128	147	164	158	175
54	128	150	165	160	177
55	129	150	165	160	177
55	132	151	168	160	178
	132	151	168	160	178
	133	151	168	160	180
		151	168	160	180
		152	169	160	180
		152	171	160	183
		152	172	160	187
		152	174	160	188
		153	176	160	188
		153		160	
		155		160	
		155		162	
		156		162	
		156		162	
		157		163	
		158		163	
		171		163	
				165	
				165	
				165	
				165	
				165	
				165	
				165	
				165	
				165	
				165	
				165	

Height Data, Sheet 2

Heights of Basketball Players (centimeters)		
NBA All-Star Game Players (1989)	Charlotte Hornets (1992–93)	Chicago Bulls (1992–93)
180	158	185
183	180	185
185	192	185
190	192	195
192	192	195
195	192	198
195	198	200
197	198	205
198	200	205
198	202	208
200	205	210
200	207	213
202		
203		
205		
205		
205		
205		
207		
210		
210		
210		
215		
220		

INVESTIGATION 24
Raisins Revisited, Revisited

Overview

In this activity, information from Investigation 22, *Raisins Revisited*, provides a context for participants to gain additional experience in making and interpreting box plots.

Reference

Friel, S. N., J. R. Mokros, and S. J. Russell. *Statistics: Middles, Means, and In-Betweens*. Palo Alto, California: Dale Seymour Publications, 1992.

Developing the Activity

For this activity you will need to return to data collected in Investigation 22. You may want to make your own set of box plots to display for the whole group to see.

Pose the Question

We collected data about different brands of raisins earlier and displayed it in a variety of ways, describing and comparing our findings for the shape of the data for each brand. Now we want to be more detailed in our comparisons of the different brands.

Collect the Data

You may want to begin a discussion by using the transparencies, which show samples of raisin data and two box plots. When you display Transparency 24.b, ask participants which brand goes with which box plot.

Using this previous data, make box plots for each brand of raisins.

Materials

Calculators

Transparencies 24.a–24.b

Handout 24.1

Analyze the Data

As a group, discuss what participants know about the different brands of raisins from the box plots. Focus attention on the box (which represents 50 percent of the data), the whiskers, and where the median falls within each box. Compare locations of these parts of the different box plots.

For example, in Transparency 24.a, 75 percent of the data for Brand A contains more raisins than 75 percent of the data for Brand B. The median number of raisins for Brand B is almost the minimum number of raisins for Brand A (with the exception of one outlier). The boxes are the same length. The spread in the data for Brand A is greater than the spread for Brand B. Questions that might be raised include the following:

Are there any differences in the weight of the raisins in each data set? Is one brand typically more plump than another?

Interpret the Results

Pay attention to what additional questions emerge as a result of this investigation and what further information participants have gained in their exploration of box plots.

Summary

Participants may be interested in exploring the question of the number of raisins in a box and the weight of the raisins in a box. If so, they can use Handout 24.1, which contains information on weight from two samples. They will need to make box plots for both the number of raisins and the weight, then make comparisons between the two brands.

If participants are curious about regulations concerning packaging and sale of goods by weight, they may want to talk to their supermarkets about their questions.

When scatter plots are introduced in Investigation 41, this data may be revisited, showing both brands—weight and number of raisins—on a scatter plot using two different colors.

Raisin Data Line Plots

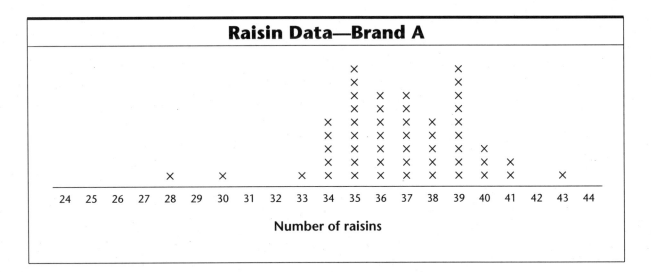

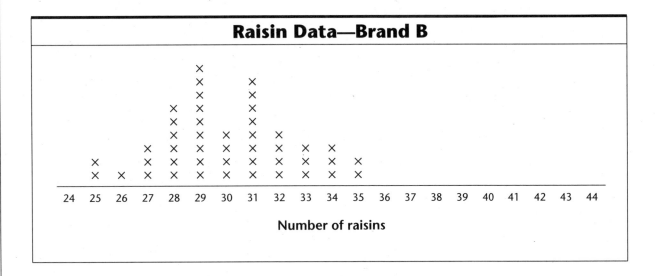

Reference

Friel, S. N., J. R. Mokros, and S. J. Russell, *Statistics: Middles, Means, and In-Betweens*. Palo Alto, California: Dale Seymour Publications, 1992.

Raisin Data Box Plots

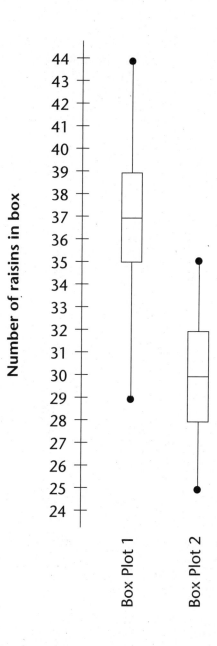

Two Brands Data

Sunmaid Raisins		Purity Supreme Raisins	
Number	Weight (g)	Number	Weight (g)
29	14.9	25	14.1
35	16.0	26	16.7
35	16.6	27	15.4
35	15.6	27	16.7
36	17.0	27	16.0
36	16.4	28	17.4
36	16.3	28	16.4
36	16.9	28	16.5
37	17.3	28	15.6
37	15.6	28	15.3
38	16.3	29	16.8
38	16.3	29	16.2
38	16.3	29	16.4
38	16.9	29	17.1
38	17.3	29	16.6
38	17.6	29	16.2
39	17.8	29	15.9
39	16.7	30	16.7
39	18.4	30	16.4
39	16.6	30	16.3
40	16.3	31	15.9
40	17.9	31	15.7
40	17.4	31	16.4
40	16.2	31	16.9
40	16.8	31	16.1
40	17.4	31	16.6
40	17.1	32	16.6
41	17.4	33	16.6
41	16.6	33	17.1
41	16.6	34	16.9
42	17.0	34	18.1
42	17.5	35	17.6
44	18.5	35	17.3

INVESTIGATION 25
Graphing Data Using Computers

Overview

If you have access to only one computer, do this investigation as a demonstration. If you can get lab sets of computers, the activity can be framed in a more investigative way.

Developing the Activity

This activity will give participants an overview of how computers can help in displaying data.

Choose at least two different data sets for participants to consider, such as data from several brands of raisins, data from different brands of cereal, or height data from Handout 23.1. Enter the data of the two sets into a computer using a statistical software program. Make a number of different graphs of the data, including box plots.

Have participants use the data you have entered to explore the different kinds of graphs that can be made with the software. If participants have access to computers, have them enter data from one of the data sets and create various graphs.

Discuss the graphs and what participants know about the data as they examine the graphs.

Discuss ways in which computers can facilitate students' work. The dynamic nature of software makes the what-if process quite accessible to students.

Note the importance of being able to adjust scale and of being able to try different representations.

Materials

Computer(s) and statistical software, data sets (such as data on various brands of raisins, brands of cereals, or height data from Handout 23.1), calculators

Handouts 25.1, 25.2

Sunburst Communications. *Data Insights* and *Statistics Workshop*. Pleasantville, New York.

MECC. *MacStat*. Minneapolis, Minnesota.

Addison-Wesley. *Data Wonder*. Menlo Park, California.

Raisins in a Half-Ounce Box

Star	Sunmaid		Purity		Del Monte		Dole	A&P
30	20	38	26	33	28	35	36	33
32	25	38	26	34	32	33	32	35
32	30	38	26	35	26	33	29	37
33	31	38	28	30	28	33	34	32
34	31	38	28	36	31	33	35	34
34	31	38	28	35	26	33	33	38
35	32	38	28	36	29	33	31	36
35	32	39	28	34	28	33	30	33
35	32	39	29	31	28	33	34	32
35	33	39	29	35	28	33	30	33
35	33	39	30	30	29	33	34	36
35	33	39	30	37	29	33	32	37
35	34	39	30	32	29	33	33	
35	35	39	30	31	29	33	30	
35	35	39	30	35	29	33	31	
36	35	39	30	32	29		31	
37	35	39	31	34	30		32	
37	35	39	31	36	30			
37	35	40	32	23	30			
37	36	40	32	27	30			
37	36	40	32	29	30			
39	36	40	32	28	30			
39	36	40	32	30	30			
39	36	40	33	27	30			
40	36	41	33	27	30			
40	36	41	34	27	31			
41	36	41	28		31			
	36	41	28		31			
	37	41	26		31			
	37	41	25		31			
	37	41	29		31			
	37	41	29		31			
	37	42	29		31			
	37	42	30		32			
	37	42	25		32			
	37	43	23		32			
	37	43	25		32			
	37	43	24		32			
	37	43	26		34			
	38	43	32		34			
	38	43	31		34			
	38	43	29		34			
	38	43	31		34			
	38	44	30		35			
	38	44	35		35			
	38	44	36		35			
	38	45	35		35			
	38	45	34		35			
	38	45	35		35			

Cereal Data

	Breakfast Cereal	Calories	Sodium (mg)	Sugars (g)	Shelf Location
1	100% Bran	70	130	6	3
2	100% Natural Bran	120	15	8	3
3	All-Bran	70	260	5	3
4	All-Bran with Extra Fiber	50	140	0	3
5	Almond Delight	110	200	8	3
6	Apple Cinnamon Cheerios	110	180	10	1
7	Apple Jacks	110	125	14	2
8	Basic 4	130	210	8	3
9	Bran Chex	90	200	6	1
10	Bran Flakes	90	210	5	3
11	Cap'n Crunch	120	220	12	2
12	Cheerios	110	290	1	1
13	Cinnamon Toast Crunch	120	210	9	2
14	Clusters	110	140	7	3
15	Cocoa Puffs	110	180	13	2
16	Corn Chex	110	280	3	1
17	Corn Flakes	100	290	2	1
18	Corn Pops	110	90	12	2
19	Count Chocula	110	180	13	2
20	Cracklin' Oat Bran	110	140	7	3
21	Cream of Wheat (Quick)	100	80	0	2
22	Crispix	110	220	3	3
23	Crispy Wheat & Raisins	100	140	10	3
24	Double Chex	100	190	5	3
25	Froot Loops	110	125	13	2
26	Frosted Flakes	110	200	11	1
27	Frosted Mini-Wheats	100	0	7	2
28	Fruit & Fibre Dates, Walnuts, Oats	120	160	10	3
29	Fruitful Bran	120	240	12	3
30	Fruity Pebbles	120	135	12	2
31	Golden Crisp	110	45	15	1
32	Golden Grahams	100	280	9	2
33	Grape-Nuts Flakes	110	140	5	3
34	Grape-Nuts	100	170	3	3
35	Great Grains Pecan	110	75	4	3
36	Honey Grahm Oh's	120	220	11	2
37	Honey Nut Cheerios	120	250	10	1
38	Honey-Comb	110	180	11	1
39	Just Right Crunchy Nuggets	110	170	6	3

3 = top
2 = middle
1 = bottom

HANDOUT 25.2

	Breakfast Cereal	Calories	Sodium (mg)	Sugars (g)	Shelf Location
40	Just Right Fruit & Nut	140	170	9	3
41	Kix	110	260	3	2
42	Life	100	150	6	2
43	Lucky Charms	110	180	12	2
44	Maypo	100	0	3	2
45	Mueslix Raisins, Dates, Almonds	150	95	11	3
46	Mueslix Raisins, Peaches, Pecans	150	150	11	3
47	Mueslix Crispy Blend	160	150	13	3
48	Multi-Grain Cheerios	100	220	6	1
49	Nut & Honey Crunch	120	190	9	2
50	Nutri-Grain Almond-Raisin	140	220	7	3
51	Nutri-Grain Wheat	90	170	2	3
52	Oatmeal Raisin Crisp	130	170	2	3
53	Post Natural Raisin Bran	120	200	14	3
54	Product 19	100	320	3	3
55	Puffed Rice	50	0	0	3
56	Puffed Wheat	50	0	0	3
57	Quaker Oat Squares	100	135	6	3
58	Quaker Oatmeal	100	0	0	1
59	Raisin Bran	120	210	12	2
60	Raisin Nut Bran	100	140	8	3
61	Raisin Squares	90	0	6	3
62	Rice Chex	110	240	2	1
63	Rice Krispies	110	290	3	1
64	Shredded Wheat	80	0	0	1
65	Shredded Wheat 'n' Bran	90	0	0	1
66	Shredded Wheat Spoon Size	90	0	0	1
67	Smacks	110	70	15	2
68	Special K	110	230	3	1
69	Strawberry Fruit Wheats	90	15	5	2
70	Total Corn Flakes	110	200	3	3
71	Total Raisin Bran	140	190	14	3
72	Total Whole Grain	100	200	3	3
73	Triples	110	250	3	3
74	Trix	110	140	12	2
75	Wheat Chex	100	230	3	1
76	Wheaties	100	200	3	1
77	Wheaties Honey-Gold	110	200	8	1

3 = top
2 = middle
1 = bottom

INVESTIGATION 26

Family Size Revisited

Materials

Stick-on notes, graph paper, colored markers, interlocking cubes in 8 colors, calculators

Transparencies 26.a, 26.b

Handouts 26.1, 26.2

Overview

Participants explore building a model for the concept of the mean. They begin by "evening out" data values to find the mean. This technique is linked to "balancing a distribution," the idea that any deviation from the mean in one direction must be matched, or balanced, by an equal deviation in the opposite direction.

Assumptions

Participants have previously considered and collected data for the problem of the typical family size in Investigation 13, *Family Size*. They know how to compute the mean and median of a data set.

Goals

Participants gain an understanding of the mean as a measure of center. In particular, they

- describe data by estimating the mean

- construct sets of data for a given mean

- understand the mean as a way of evening out a set of data and as a balance point in a set of data

- understand that the mean alone may not give enough information about a set of data

References

Friel, S. N., J. R. Mokros, and S. J. Russell. *Statistics: Middles, Means, and In-Betweens*. Palo Alto, California: Dale Seymour Publications, 1992.

Fey, J., W. Fitzgerald, S. Friel, G. Lappan, and E. Phillips. *Data About Us*. Palo Alto, California: Dale Seymour Publications, Forthcoming.

Developing the Activity

Participants will explore representations of the mean using towers made of interlocking cubes and line plots made of stick-on notes.

Pose the Question

Remind participants of their work in Investigation 13, *Family Size*. Ask them how they defined a family in that investigation.

Discuss examples of averages from school, the community, the news, and the popular media.

Remember that an average is a single number or value often used to describe what is typical about a set of data. It can be thought of as a "measure of location." The median is one kind of average that we have used quite a bit. It shows us the location at which a set of ordered data is divided in half: half the data are below that point, and half the data are above it. Now you will explore another kind of average, the mean.

Pose this problem for participants to consider.

Eight students in one class gathered data about the number of people in their families. Each student made a tower of cubes to show the number of people in his or her family. Here is what they found.

As you describe the problem situation, display eight towers of interlocking cubes. Use a different color for each tower. The first letter of each name suggests a color to use (such as Yvonne for yellow; Brendon, Batai, and Beth for black, dark blue, and light blue).

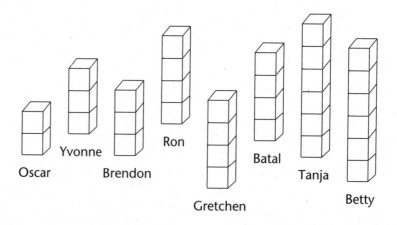

Work with participants to see the frequencies of data values implicit in the cube towers and then make a line plot with stick-on notes to show the data.

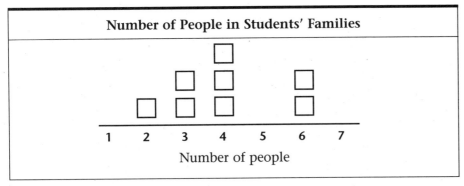

Discuss how many people there are altogether in the eight families. Make sure participants can obtain this information from both the cube towers and the line plot. Explain that students may not immediately see that the cube towers and the line plot are representations of the same information.

Analyze the Data

Ordering the data towers enables us to group by number of people.

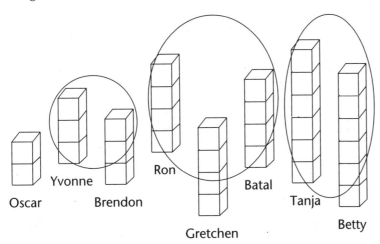

The students wondered how many people there are in the eight families on average. What is the mode family size for these eight families? How is it shown with the cubes? How is it shown with the stick-on note line plot? Do they show the same mode? Why might students be confused with what the mode is?

Students often think of the mode as the *most* in a data set and miss the subtle distinction needed to understand that the mode is the *most frequent* data value. It is not unusual for a student to say something such as, "The mode in the towers is 6 and the mode in the line plot is 4." This reflects students' lack of connection between the two representations.

What is the median family size for these eight families? How can you find it using the cube towers? How else can it be found?

The strategy students will most likely use with the cube towers is to work in from the ends: pairing the shortest and tallest towers, removing each pair until they reach the middle two data values, then computing the median.

The students found another way to find an average: by moving people to other families to "even out" the sizes of the families. They know they must keep a total of 32 people in the eight families.

With participants, add the cubes that are in all the towers.

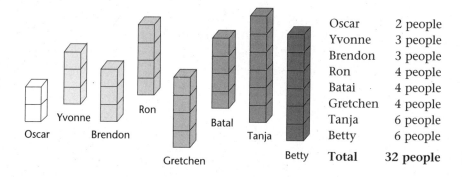

Oscar	2 people
Yvonne	3 people
Brendon	3 people
Ron	4 people
Batai	4 people
Gretchen	4 people
Tanja	6 people
Betty	6 people
Total	**32 people**

The students moved cubes from one tower to another, making some families bigger than they actually were and other families smaller than they actually were. When they were finished, their cube towers looked like this.

Move cubes from taller towers to shorter towers until every tower has four cubes.

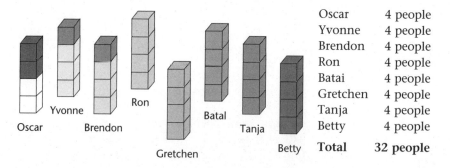

Oscar	4 people
Yvonne	4 people
Brendon	4 people
Ron	4 people
Batai	4 people
Gretchen	4 people
Tanja	4 people
Betty	4 people
Total	**32 people**

Because you used different colors for each family, participants will be able to see how the evening out is accomplished. You may want to highlight the number of people in each family as shown both before and after the evening-out process, and the total number of people in all the families.

This representation shows another kind of middle number. We can say that, on average, there are 4 people in each family. We call this number the **mean**. *The mean number of people in the eight families is 4. There are a total of 32 people in the eight families.*

Display Transparency 26.a, which shows the line plot the students made for the eight families and depicts the mean as a balance point. Distribute Handout 26.1, and discuss the questions posed on it.

Notice that the mean is shown as a location on the number line and is indicated by means of an arrow. The mean is a kind of balance point. We can show that many family sizes are not the same as the mean but are related to the mean. Some are less than the mean, and some are more than the mean.

Another group of eight students in the class had the following data about the number of people in their families.

Again, make eight cube towers, using distinctive colors for each. Then make a line plot to show this new set of data.

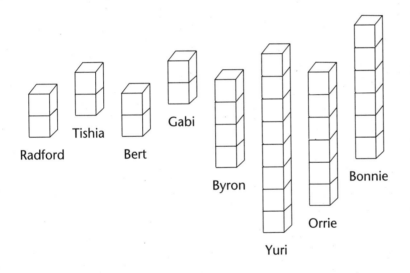

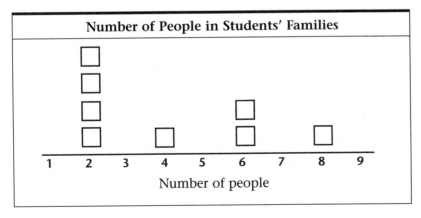

Family Size Revisited **249**

What is the mean number of people in these eight families? How can we show a balance point for the data displayed on the graph?

Display Transparency 26.b. Again, point out that the sum of the differences from the mean for data values below the mean is the same as the sum of the differences from the mean for data values above the mean.

Have participants work in small groups to consider the questions on Handout 26.2.

Interpret the Results

It is possible to create other sets of eight families with a mean of 4 people. Working with a partner, see if you can find two more sets of eight families with a mean of 4 people. Use cubes to show each set, then make line plots that show the information from the cubes.

Distribute Handout 26.3, which will give participants more experience with these concepts.

Summary

This is the first part of our introduction to the mean. The most important outcome is flexibility in moving between the representations of the data and being able to talk about the mean and how it can be pictured.

The Mean as a Balance Point: Data Set 1

Number of People in Students' Families

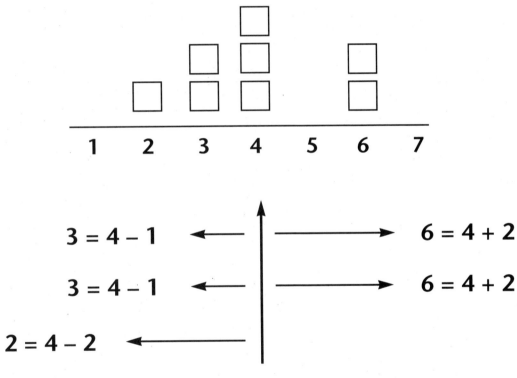

The Mean as a Balance Point: Data Set 2

Number of People in Students' Families

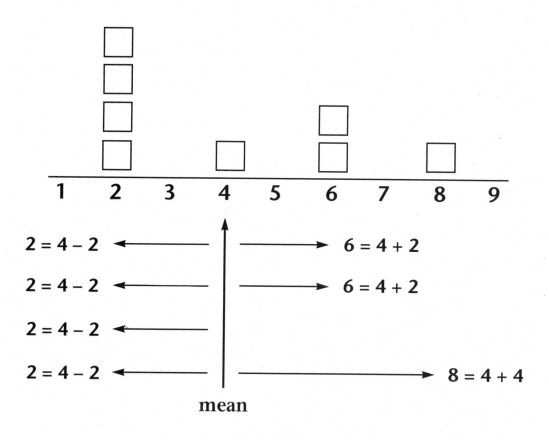

The Mean as a Balance Point

Look at the line plot below.

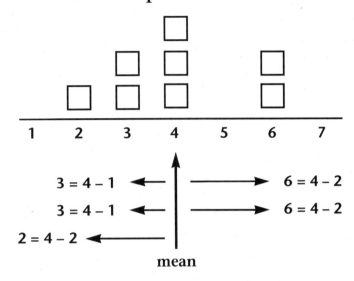

1. Which families have the same number of people as the mean number of people in a family?

2. Which families have fewer people than the mean number of people? How many *fewer* people than the mean are in each of these families? What is the total number of "fewer people than the mean" for these families?

3. Which families have more people than the mean number of people? How many *more* people than the mean are in each of these families? What is the total number of "more people than the mean" for these families?

4. What do you notice about the total number of "fewer people" and the total number of "more people" that you found in questions 2 and 3? Explain what you discover.

These questions highlight the deviations from the mean. The differences of the data greater than the mean and the differences of the data less than the mean are the same. The mean *balances* the distribution.

Comparing Balance Points

Look at the two line plots below. The data for the two sets of eight families look different, but the mean is the same.

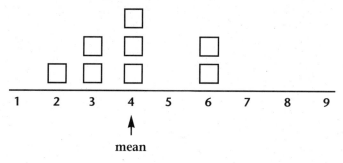

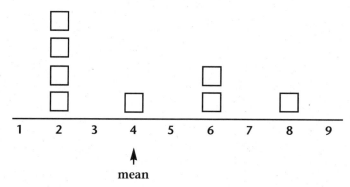

1. How many total people are represented in each situation? How does this relate to the mean being 4 in each case?

2. a. Are the data in one line plot more spread out than the data in the other line plot?

   b. How does knowing the range of the data help you answer this question?

   c. How does knowing the total number of "fewer people than the mean" and the total number of "more people than the mean" help you answer this question?

More Work with the Mean

1. A group of 8 students has a mean number of 3 people in their families. Make a representation with cubes that shows these 8 families. Then make a line plot, using an arrow to show the mean as a location on the line plot. (Hint: If there are 8 families with a mean of 3 people in each, how many people are there in all?)

2. A group of 10 students has a mean number of 3 people in their families. Make a representation with cubes that shows these 10 families. Then make a line plot, using an arrow to show the mean as a location on the line plot.

3. A group of 10 students has a mean number of 3½ people in their families. Make a representation with cubes that shows these 10 families. Then make a line plot, using an arrow to show the mean as a location on the line plot.

   You may want to discuss with others how you can have a mean of 3½ people, when we know families really can't have fractions of people in them!

4. What is the mean number of people in your participants' families? Explain how you can determine this using interlocking cubes. Make a line plot of the data, showing the mean as a balance point. Discuss why the mean is a balance point.

INVESTIGATION 27

Cats

Overview

Participants work with a data set about 24 cats to determine whether they can decide from the sample what is a "typical" cat. They organize, display, and summarize the data. Following Part 1, each participant collects data on one more cat to expand the sample. In Part 2, participants analyze the new data to decide whether they want to change their descriptions of a typical cat.

Assumptions

Participants will be familiar with a variety of ways to display data and to determine what is "typical" about a data set. They have considered ideas related to samples. Some instruction and discussion will probably be needed on representativeness of samples and biased samples.

Goals

Participants explore the concept of sampling. In particular, they

- understand the concept of random sampling

- learn about different types of data that may be collected

- examine the results from the entire investigation

- recognize that different interpretations of the same set of data are possible

References

Corwin, R. B., and S. N. Friel. *Statistics: Prediction and Sampling.* Palo Alto, California: Dale Seymour Publications, 1990.

Taylor, D. *You and Your Cat.* Westminster, Maryland: Random House, 1986.

Materials

Statistics: Prediction and Sampling for the sets of 24 cat cards, graph paper, scissors, colored markers, calculators, colored stick-on dots, chart paper, tape

Handout 27.1

Fun Facts

A cat's pupils look like slits during the day and become round at night.

Tabby may have night vision up to six times better than her owner's vision.

A cat's whiskers are so sensitive that they can sense the currents of air around objects when the cat is slipping about in the dark of night.

Developing the Activity: Part 1

Participants will examine a data set and analyze what is typical in the data.

Pose the Question

Have participants listen to the following information and see whether they can decide what animal is being described.

An average adult is about 12 inches high at the shoulder and 18 inches from head to tail base, has a tail that is about 12 inches long, and weighs between 6 and 12 pounds (Taylor 1986).

Stop for a few minutes, and give participants a chance to recognize that you are talking about cats.

We can't possibly examine all cats in the population, so we will instead examine a set of cards with information about a small sample of cats.

Assign the participants to teams of two to three, and hand out the card packs, one pack to a team. Allow about 5 minutes for them to browse through the cards, making observations and asking questions.

You are going to decide what a "typical" cat in this sample would look like. First, you will research one characteristic of the cats. Decide what characteristic your team wants to investigate.

Let teams talk about this. On the board, record which characteristics will be investigated. Make sure all of the attributes are chosen so that participants will get a complete picture of a cat.

Collect and Analyze the Data

Take about 20 minutes to decide what is typical of this sample of cats. When you report back, you will need data to support your conjectures. Some kind of display or graph will help you to present your findings.

Encourage each team to make observations about the shape of the data. Ask them to discuss which measure of center is most appropriate and why it is appropriate.

Interpret the Results

As teams report their findings to the group, keep track of the findings. Encourage the group to ask questions about each team's findings.

How well does this sample approximate the characteristics of all cats? Are there any colors or kinds of cats that are missing?

Summary

Distribute Handout 27.1 to each participant. Ask them to collect data on at least one cat, either their own or a friend's or neighbor's. They will use their data at the next session.

Developing the Activity: Part 2

Participants will add new data to the set of data they investigated in Part 1. Post chart paper and have each participant fill in columns with the data each collected: name, gender, age, weight, and so on.

Pose the Question

Today you will combine the cat data you gathered with the data from the original 24 cats. You will record the data in the appropriate places on the sheets that are taped to the wall around the room. Then you will get together in your teams from before and analyze one of the characteristics.

Collect the Data

Participants record the data they collected on the sheets that are taped to the walls.

Analyze the Data

Let participants decide which characteristic they would like to analyze, but ask them to choose a different characteristic from the one they looked at previously. Make sure all the characteristics are covered.

Ask each team to create a graph to help present its findings. Encourage some teams to use box plots. Ask them to be prepared to explain which measures of center they chose to use and why.

Teacher Notes

Participants may notice that there are no cats with blue eyes or white coats. Is this a representative sample of the cat population? Talk about ways in which the sample is representative and ways in which it is not. What might participants do to get a better sample?

Interpret the Results

As teams report their findings to the group, keep track of them on a piece of chart paper. Encourage the class to discuss the methods teams used to make their decisions, and talk about different methods of determining typical characteristics.

Discuss the relationship between the size of a sample and how well it might approximate the population. Note that it could be the case that none of the cats in the sample matches the description of a typical cat.

Fun Facts

Not only can a cat's tongue curl backward into a little cup shape to lap up liquids, but it is serrated with small spikes that help in cleaning and eating.

HANDOUT 27.1

Cat Data Collection Sheet

Cat's name		
Gender	Age (years)	Weight (pounds)
Body length (inches)		Tail length (inches)
Fur color	Eye color	Pad color
Other		

Cat's name		
Gender	Age (years)	Weight (pounds)
Body length (inches)		Tail length (inches)
Fur color	Eye color	Pad color
Other		

Reference

Corwin, R. B., and S. N. Friel. *Statistics: Prediction and Sampling.* Palo Alto, California: Dale Seymour Publications, 1990; p. 91.

INVESTIGATION 28

Technology

Overview

Computers are invaluable tools for organizing, sorting, searching, and manipulating data. These technology-based activities will acquaint participants with software that can help them organize and display data in their classrooms.

How this investigation unfolds depends on what types of computers and what graphing and database software is available for participants to explore. Offer them experience with as many types of computers and software packages as possible.

The power of the computer can easily be demonstrated with large data sets. You will have to prepare the data files in advance by entering the information for two different data sets. The first data file, which you might call CATS, will contain the information about cats described in *Statistics: Prediction and Sampling,* on the Cat Cards. The second data file, which you might call NEWCATS, will contain the information about cats that participants gathered in Investigation 27, *Cats.*

Assumptions

Participants gain experience with database and graphing software. Before they can effectively use technology in their classrooms, they need experience with appropriate technology working as students. Participants may have no experience with database and graphing software and need an introduction to the structure and use of the materials. They may or may not be familiar with specific computers.

Goals

Participants explore the use of technology to analyze statistics. In particular, they

- understand that technology provides powerful tools for working with data

- recognize that database and graphing software can be used to represent data in a variety of ways and provides an opportunity to look at data holistically

Materials

For each lab: Computer (such as an IBM PC or compatible, Apple II, or Macintosh computer), database and graphing software (such as *Data Insights* or *Statistics Workshop,* available through Sunburst Communications, *MacStat* available through MECC, or *Data Wonder* available through Dale Seymour Publications), data disks of 2 sets of data; copy of *Statistics: Prediction and Sampling*

Handouts 28.1 through 28.3

Reference

Corwin, R. B., and S. F. Friel. *Statistics: Prediction and Sampling.* Palo Alto, California: Dale Seymour Publications, 1990.

Developing the Activity

Form teams of two or three, depending upon the available computer facilities. If several labs are used, participants may be divided into groups and rotate through the labs.

Browsing, Sorting, and Searching

Show participants how to start the database program, create new data disks (if necessary), and retrieve a file that has already been created.

Begin work with the file containing the original cat data. Discuss the terms *records* and *fields* and the options on the main menu. Show participants how to enter a new record. Let them practice with the data for Wally given on Handout 28.1. Show them how to browse through the database, how to sort the data by a particular characteristic, how to order the data, how to search for a particular set of cats, and how to print reports.

Distribute Handouts 28.2 and 28.3, and circulate as participants answer the questions. If there is time, ask them to answer the same questions using the file containing their cat data.

Merging and Summarizing

In Investigation 27, *Cats,* participants analyzed data from 24 cat cards to come up with a description of a typical cat and then analyzed data that they had collected. Now have them merge the two sets of data using the computer and then examine the characteristics of a "typical" cat one last time. They can use graphs or line plots when necessary in their analysis.

Compare your three descriptions of a "typical" cat. How are they different? How are they alike? Do we have a large enough sample to describe a typical cat? How many cats should you measure to describe a typical cat?

Discuss how participants would use database and graphing software in their classrooms. If you have time, you may want to have participants explore other database files—either data sets that you or they enter, or files that were packaged with the database software.

Exploring Cats: Entering Records

You can add new records to a database at any time to keep your file up to date.

Add Wally as a new record.

Cat's name		
Wally		
Gender	Age (years)	Weight (pounds)
male	5 years	10 pounds
Body length (inches)		Tail length (inches)
18 inches		12 inches
Fur color	Eye color	Pad color
black and white	green	pink and black
Other		
Wally is Peeble's brother.		

HANDOUT 28.2

Exploring Cats: Finding Records

You can find records in a database at any time. The *Search/Find* function is simply a way to search the database to locate particular files.

Use the *Search/Find* function to find *your* cat. Once your cat is found, answer these questions.

1. Which cats have pink pads?

2. How many cats have black fur?

3. How many cats have green eyes?

4. How many male cats are more than 5 years old?

Exploring Cats: Sorting Records

You can sort records in a database at any time. Sorting is organizing files in a particular way. For example, files can be organized in alphabetical order, by gender, or by age.

Use the *Sort* function to sort the records alphabetically by cat name and answer these questions.

1. Which is the last cat in alphabetical order?

2. Which cat has the shortest tail?

3. Which cat is heaviest?

4. How many cats are older than your cat?

5. Which cats are old and fairly heavy?

INVESTIGATION 29
Building the "Rule" for Finding the Mean

Overview

Participants develop the algorithm for finding the mean. This development is built on their earlier work with family size in Investigation 26, *Family Size Revisited*.

Assumptions

Participants are familiar with the process of data investigation, have been involved in several investigations involving measures of center (mode and median) and graphical representations for analyzing data, and have completed the *Family Size Revisited* investigation.

Goals

Participants formalize the procedure for identifying the mean. In particular, they

- understand the algorithm for computing the mean

- understand the role of zero in determining the mean of a data set

- explain the impact on the mean of very high or very low values

Reference

Fey, J., W. Fitzgerald, S. Friel, G. Lappan, and E. Phillips. *Data About Us*. Palo Alto, California: Dale Seymour Publications. Forthcoming.

Developing the Activity

Participants will examine a new set of data and explore strategies for finding the mean of the data set.

Materials

Interlocking cubes, calculators

Handouts 29.1, 29.2

Teacher Notes

Students who answered the survey may have been confused about what was meant by "movies." Some may have thought this meant only movies seen in a theater. Others may have thought it included movies on video. Still others may have included movies seen on television.

Pose the Question

Discuss with participants that they may often encounter much larger data sets than they have been considering; instead of 8 families, they may have data from 100 families.

The methods you have been exploring for finding the mean would be difficult to employ with a large data set; we need another way to find the mean. You may already know the rule, but this investigation may help you explain why the rule works.

Introduce the data set shown on Handout 29.1.

In a recent survey of middle-grades students, one of the questions they were asked to respond to was, "How many movies do you watch in one month?"

How would you answer this question? What kinds of things must you decide before you can answer?

Collect the Data

Distribute Handout 29.1.

The data that were gathered are shown on the handout. Notice that the data are quite spread out; the range is from 1 to 30 movies.

Analyze the Data

Surveys like this one are conducted to help us understand what is typical for a particular group. We can use the mean to help us describe what's typical.

If each student had watched the same number of movies in one month, how many movies would that be? Describe what you need to do to answer this question.

Let participants explore this question in small groups. If groups want assistance, you might suggest that they first consider only the data from the first five students.

Gender	Movies per month
Boy	2
Boy	15
Girl	13
Girl	1
Boy	9

What is the total number of movies watched by these five students?

How could you divide the movies among the five students so that each student will have watched the same number of movies?

Make a stem plot of these data.

Next, suggest that participants look at the data from all the students.

What is the total number movies watched by all the students?

How could you divide the movies among all the students so that each student will have watched the same number of movies?

Add the remaining students' data to your stem plot.

Have a discussion about the strategies participants used to answer the question, "If each student had watched the same number of movies, how many movies would that be?" What you want to emerge is the discovery of the algorithm of "add them all up and divide by the number of numbers." Using their earlier experiences with evening out and finding the balance point, participants should have reference points that might help them to justify the algorithm.

When the discussion has concluded, distribute Handout 29.2, which contains additional problems designed to highlight issues that often arise in working with the mean.

Survey Results

Movies Watched by Students in One Month

Gender	Movies per month
Boy	2
Boy	15
Girl	13
Girl	1
Boy	9
Girl	30
Boy	20
Boy	1
Girl	25
Girl	4
Girl	3
Boy	2
Boy	3
Boy	10
Girl	15
Boy	12
Boy	5
Boy	2
Girl	4
Boy	1
Girl	4
Boy	11
Boy	8
Boy	5
Girl	17

Mean Problem Sheet

1. A new student joins the class that was surveyed. When asked how many movies she watched last month, she said, "None." This means that we add this data:

    Girl   0

    Does this change the mean number of movies watched for the class? Explain your reasoning.

2. Suppose there were 50 students in the data set. What would you need to know to determine the mean number of movies watched by these students?

3. If the mean number of movies watched was 15, what would you need to know to determine how many movies were watched in all?

4. If the total number of movies watched was 630, what would you need to know to determine the number of students surveyed?

Now, add these data to the results of the survey of middle-grades students about movies watched in one month.

Boy	15
Girl	16
Girl	5
Boy	18
Girl	3
Girl	6
Girl	7
Boy	6
Girl	3
Boy	11

HANDOUT 29.2

5. Compute these values. You may want to use your calculator.

Total number of students surveyed:

Total number of movies watched:

Mean number of movies watched:

6. Another student's data is added.

                Girl   42

Compare the mean you found in question 5 with the mean you find now. What do you notice? Explain.

7. Another student is surveyed.

                Girl   96

Compare the mean you found in questions 5 and 6 with the mean you find now. What do you notice? Explain.

8. Eight more students' data are added.

                Boy   5
                Girl   5
                Boy   5
                Girl   5
                Girl   4
                Girl   4
                Girl   2
                Boy   2

Compare the mean you found in questions 5, 6, and 7 with the mean you find now. What do you notice? Explain.

INVESTIGATION 30

Means in the News

Materials

Colored markers, chart paper, colored stick-on dots, calculators

Handout 30.1

Overview

In Investigation 26, *Family Size Revisited,* participants studied data sets that had the same mean or median but looked very different. Discussing those examples and working through this activity will help them to recognize that the mean or median alone conveys only limited information. Activities such as this one are appropriate for assessing students' understanding of the concept of average.

Assumptions

Participants know how to compute the mean and understand the algorithm for computing it. They will develop a further understanding of what information the mean does and does not provide.

Goals

Participants consider the use of the mean to communicate information. In particular, they

- understand the appropriateness of various types of representations and identify misrepresentations of data

- use various measures of center to describe and compare distributions

Reference

Friel, S. N., J. R. Mokros, and S. J. Russell. *Statistics: Middles, Means, and In-Betweens.* Palo Alto, California: Dale Seymour Publications, 1992.

Developing the Activity

Participants will consider how much information a single statistic—such as the mean—conveys, then write a report analyzing the information actually conveyed in a single published statistic.

Pose the Question

When we worked through the problems in some of the previous investigations, we sometimes found that data sets that look quite different can have the same mean or median. Do you remember any of those situations?

Discuss briefly what they remember about how an average could be the same for different sets of data.

We are going to look at some more averages and think about what those averages tell us and what they don't tell us. You will often be given an average and not be able to view the entire set of data. For example, in 1990 Ben and Jerry's ice cream company reported that the average amount of ice cream eaten by a person in the United States was 45.14 pints per year.

What does this average tell us? What might the data have looked like?

As the group discusses these questions, sketch a couple of graphs to accompany what they are describing. Then have them work in pairs to create two sketches that look very different yet agree with the average given by the company.

Post participants' graphs, and discuss what kind of situation each might represent. Ask participants which graphs are the most realistic.

You can see that just knowing the mean or the median is not enough to tell you what the data set looks like. It would be best if, when people report data and statistics, they also relay information to help establish the circumstances under which the data were collected, but often they do not.

Now you will use what you have learned to interpret some real averages that were reported in newspapers and other sources.

Collect and Analyze the Data

Distribute Handout 30.1. Have each pair of participants select one average and write a brief report about what the average tells about the data. Participants may include graphs if they feel they would be helpful. They may also comment on what information they believe is missing. Emphasize that participants can use personal knowledge and experience to help them with the task.

Interpret the Results

Have pairs present their reports. Challenge them to justify their graphs.

Encourage discussion and question interpretations. Discussion should focus on what the mean does and does not communicate about the data and what additional information would help give a better picture of the data set.

Summary

We hope participants will recognize that when reporting about data, they should provide as rich a description as possible, including the mean or median, the range, and important characteristics about the spread of the data. We also hope that participants will recognize that when only one measure is given, statistical interpretation should be performed with caution.

HANDOUT 30.1

Real Means

1. In an American household, the television is on for an average of 7 hours each day.

2. The average life span of a hamster is 3 years.

3. Children ages 10–13 get an average of 10 hours of sleep each night.

4. Americans go to an average of 5 movies each year.

5. Americans produce an average of 4 pounds of garbage a day.

Reference

Friel, S. N., J. R. Mokros, and S. J. Russell. *Statistics: Middles, Means, and In-Betweens.* Palo Alto, California: Dale Seymour Publications, 1992; p. 88, Student Sheet 10.

INVESTIGATION 31
Comparing Sets of Cereal Data

Overview

Participants use what they have learned about representing data and what they already know about the mean and the median to investigate a set of data about the sugar content of breakfast cereals. Using three data sets gathered at a grocery store, they will investigate the hypothesis that cereals with higher sugar content are placed on the middle shelf, where young children are most likely to see and select them. Participants will analyze the data, construct box plots, and prepare a report comparing the three data sets.

Assumptions

Participants can recognize and describe features of data, are familiar with the concepts of the mean and the median, and can compute the mean of a data set.

Goals

Participants explore the concepts of the median, the mean, and data representation. In particular, they

- understand the concept of the median

- understand that the mean gives no information about the shape of the data

- use the mean and the median to compare sets of data

- construct a box plot

References

Friel, S. N., J. R. Mokros, and S. J. Russell. *Statistics: Middles, Means, and In-Betweens.* Palo Alto, California: Dale Seymour Publications, 1992.

"Cereal: Breakfast Food or Nutritional Supplement?" *Consumer Reports* (October 1989): 638–46.

Materials

Nutritional-information panels from many cereal boxes (3–5 per team), graph paper, colored markers, computers and graphing software capable of constructing box plots (optional), calculators

Transparencies 31.a through 31.c

Handouts 31.1 through 31.4

Developing the Activity

Participants will first analyze the nutritional-information panels, then analyze three sets of data to investigate a hypothesis about those data.

Pose the Question

Divide participants into teams of three. Distribute Handout 31.1 and three to five nutritional-information panels from cereal boxes to each team. Ask teams to find the amount of sugar per serving for each cereal and to compare that information with the sugar content of other foods from the handout. After teams have had time to examine the data, ask them to report what they found. Discuss the serving size for the cereals. Is the amount shown on the information panel a realistic serving size?

Lots of parents are concerned these days about the sugar in their children's breakfast cereal. Many complain that grocery stores display cereals with high sugar content where they are most tempting to children: on the middle shelf. This is just about eye level for most children. Do you think this claim is true?

Allow participants a few minutes for discussion.

Collect and Analyze the Data

We are going to look at real data that were collected at a grocery store in Massachusetts. The three data sets consist of the sugar content of cereals from the top shelf, middle shelf, and bottom shelf. You will work in your teams to investigate whether there are any differences in the three sets of data. First, let's all look at the bottom shelf.

Distribute Handout 31.2. The chart lists the number of grams of sugar per serving. Ask teams to organize the data; they might make a table, a line plot, or some other kind of graph. Tell them to write down all they can about the data. Circulate while the teams work.

This is a good place to review how to construct box plots: first, identify the median, upper and lower quartiles, and upper and lower extremes. Second, construct the box and the whiskers. Third, use the "one-and-a-half" rule to check for outliers.

Interpret the Results

Bring the teams together to discuss what they found. A discussion that outlines a complete description of the data will include information about the shape of the data—clumps, range, outliers, and gaps—and about the center of the data. Someone will probably have calculated the mean.

Did you estimate before calculating the mean? Did your estimation seem reasonable? What does the mean communicate about the data?

Someone else may have calculated the median.

How did you determine the median? Why are the mean and the median different? What do they communicate about the data?

Distribute Handouts 31.3 and 31.4 to each team. Ask participants to investigate this question: Do cereals on the middle shelf of this grocery store have a higher sugar content? Assign each team these tasks:

- sketch the data from all three shelves

- jot down phrases that describe each data set

- estimate, and then calculate, the mean of each data set

- find the median of each data set

- decide how the three data sets are similar and how they are different

Box plots for each data set are appropriate graphical representations for making comparisons. You might want to use Transparencies 31.a through 31.c to review the box plots and the line plots for these data sets.

Have each team write a brief report of their findings, including graphs of the data.

Summary

This activity should generate discussion about the advantages and disadvantages of the median and the mean, and when one may be more appropriate than the other. A discussion of different types of graphical representations should include how to compare data sets. Participants may be concerned about the appropriate time to teach box plots. The use of box plots is appropriate for the middle-school curriculum, not the elementary curriculum.

Variations

You may want to share the *Consumer Reports* article on the nutritional content of cereal with participants. The article may inspire additional questions that their students might investigate.

Cereal Data Line Plots

Cereals on the Top Shelf

```
                    X       X
                    X       X
                    X       X
                    X       X   X   X
            X       X   X   X   X   X
            X   X   X   X   X   X       X
X           X   X   X   X   X   X   X   X                       X
─────────────────────────────────────────────────────────────────────
0   1   2   3   4   5   6   7   8   9   10  11  12  13  14  15
                        Grams of sugar
```

Cereals on the Middle Shelf

```
                            X                   X
                            X   X               X   X   X
                            X   X       X       X   X   X
                            X   X       X       X   X   X
                            X   X   X   X   X   X   X   X
                    X   X   X   X   X   X   X   X   X   X   X   X
─────────────────────────────────────────────────────────────────────
0   1   2   3   4   5   6   7   8   9   10  11  12  13  14  15
                        Grams of sugar
```

Cereals on the Bottom Shelf

```
X
X
X           X
X           X           X
X           X   X       X   X           X   X           X
X   X   X   X           X   X           X   X       X   X
─────────────────────────────────────────────────────────────────────
0   1   2   3   4   5   6   7   8   9   10  11  12  13  14  15
                        Grams of sugar
```

Cereal Data Box Plots

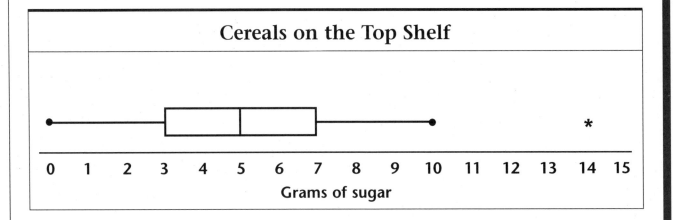

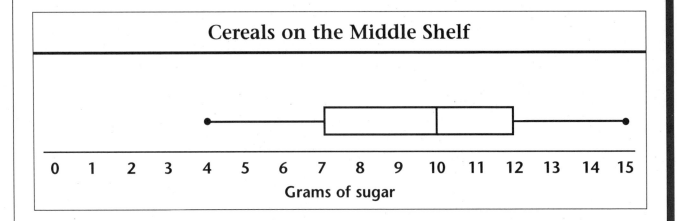

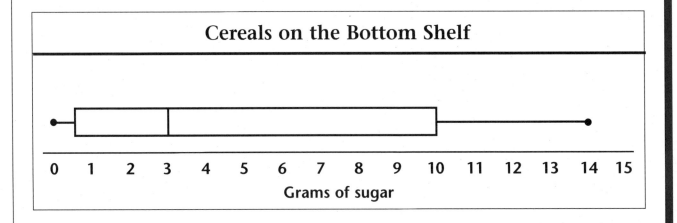

Cereal Data Line Plots and Box Plots

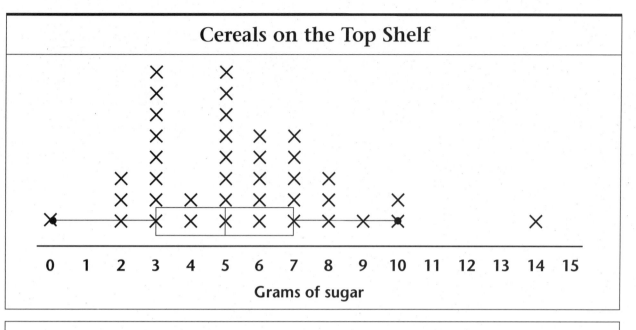

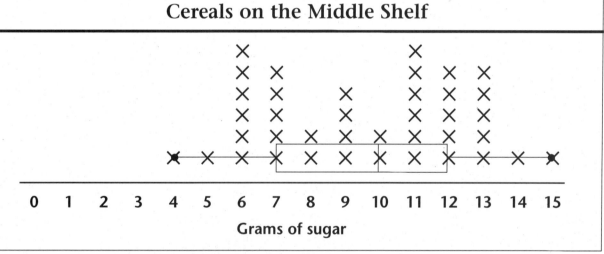

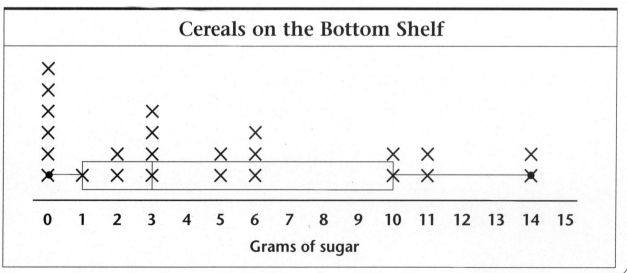

Sugar Content of Common Foods

Food	Portion (or serving)	Total sugar (grams)
American cheese	1 slice	2.1
Apple (raw)	1 medium	15.7
Bacon	2 slices	0.0
Banana (raw)	1 average	22.4
Blueberry muffin	1 average	5.7
Carrot (raw)	1 medium	4.8
Catsup	1 tbsp.	3.4
Chocolate candy bar (plain)	1 oz.	14.5
Cola	12 oz.	37.7
Cranberry juice cocktail	4 oz.	21.0
Creamsicle (orange)	1 bar	16.4
Grapefruit (raw)	½ average	4.9
Grapes (green, seedless)	½ cup	13.1
Hi-C apple drink	8 oz.	30.5
Honey	1 tbsp.	17.3
Hostess devil's food cupcake	1 cake	14.2
Ice cream, Baskin Robbins strawberry	⅔ cup	29.9
Ice cream, Baskin Robbins vanilla	⅔ cup	18.4
Milk (whole)	8 oz.	11.4
Orange (raw)	1 medium	12.1
Peanut butter	1 tbsp.	0.1
Pickle (dill, whole)	1 medium	1.8
Pudding (Jell-O instant chocolate w/whole milk)	½ cup	27.0
Raisins	¼ cup	26.9
7-Up	12 oz.	36.5
Snickers	1 oz. bar	13.3
Sugar (brown)	1 cup	216.0
Sugar (granulated)	1 cup	199.0
Sugar (white)	1 cube	7.0
Yogurt (low fat, fruit)	1 cup	43.2

Reference

Friel, S. N., J. R. Mokros, and S. J. Russell. *Statistics: Middles, Means, and In-Betweens.* Palo Alto, California: Dale Seymour Publications, 1992; p. 84, Student Sheet 6.

Cereals on the Bottom Shelf

Cereal	Sugar per serving (grams)	Brand
Super Golden Crisp	14	Post
Honey Comb	14	Post
Alphabits	11	Post
Alphabits (with marshmallow)	13	Post
Shredded Wheat	0	Sunshine
Shredded Wheat (Bite Size)	0	Sunshine
100% Bran	6	Nabisco
Shredded Wheat	0	Nabisco
Team Flakes	5	Nabisco
Shredded Wheat (Spoon Size)	0	Nabisco
Cheerios	1	General Mills
Cheerios (Apple Cinnamon)	10	General Mills
Cheerios (Honey Nut)	10	General Mills
Total	2	General Mills
Wheaties	3	General Mills
Product 19	3	Kellogg's
Rice Krispies	3	Kellogg's
Special K	3	Kellogg's
Frosted Flakes	11	Kellogg's
Corn Flakes	2	Kellogg's
Crunchy Bran	6	Quaker
Life	5	Quaker
Life (Cinnamon)	6	Quaker
Puffed Wheat	0	Quaker
Puffed Rice	0	Quaker

Reference

Friel, S. N., J. R. Mokros, and S. J. Russell. *Statistics: Middles, Means, and In-Betweens.* Palo Alto, California: Dale Seymour Publications, 1992; p. 85, Student Sheet 7.

HANDOUT 31.3

Cereals on the Middle Shelf

Cereal	Sugar per serving (grams)	Brand
Raisin Bran	5	Post
Fruity Pebbles	12	Post
Cocoa Pebbles	13	Post
Honey Bunches of Oats	6	Post
Breakfast Bears (cinnamon)	7	Post
Breakfast Bears (chocolate)	7	Post
Breakfast Bears (honey)	7	Post
Almond Delight	8	Ralston
Batman	10	Ralston
Hot Wheels	11	Ralston
Cookie Crisp (chocolate chip)	13	Ralston
Clusters	7	General Mills
Raisin Nut Bran	8	General Mills
Trix	12	General Mills
Oatmeal Crisp	6	General Mills
Golden Grahams	9	General Mills
Cocoa Puffs	9	General Mills
Cinnamon Toast Crunch	11	General Mills
Lucky Charms	11	General Mills
Count Chocula	13	General Mills
Honey Smacks	15	Kellogg's
Cocoa Krispies	11	Kellogg's
Fruitful Bran	11	Kellogg's
Frosted Mini-Wheats	6	Kellogg's
Frosted Mini-Wheats (Bite Size)	7	Kellogg's
Nut and Honey Crunch	9	Kellogg's
Just Right	9	Kellogg's
Shredded Wheat (Apple Cinnamon)	6	Kellogg's
Shredded Wheat (Raisin)	6	Kellogg's
Raisin Bran	13	Kellogg's
Apple Jacks	14	Kellogg's
Corn Pops	12	Kellogg's
Fruit Loops	13	Kellogg's
Oh's (Honey Graham)	11	Quaker
Cap'n Crunch	12	Quaker
Cap'n Crunch (Peanut Butter)	10	Quaker
Cap'n Crunch's Crunch Berries	12	Quaker
Rice Bran	6	Quaker
Oat Bran	4	Quaker

Reference

Friel, S. N., J. R. Mokros, and S. J. Russell. *Statistics: Middles, Means, and In-Betweens.* Palo Alto, California: Dale Seymour Publications, 1992; p. 86, Student Sheet 8.

HANDOUT 31.4

Cereals on the Top Shelf

Cereal	Sugar per serving (grams)	Brand
Grape Nuts	3	Post
Grape Nuts Flakes	5	Post
Raisin Grape Nuts	6	Post
Bran Flakes	5	Post
Fruit and Fiber	6	Post
Oat Flakes	6	Post
Rice Bran	7	Ralston
Fruit Muesli	4	Ralston
Chex (Multi Bran)	6	Ralston
Chex (Wheat)	3	Ralston
Chex (Oat)	5	Ralston
Chex (Rice)	2	Ralston
Batman	10	Ralston
Double Chex	5	Ralston
Total Raisin Bran	14	General Mills
Total Whole Wheat	3	General Mills
Wheaties	3	General Mills
Total Corn Flakes	2	General Mills
Kix	3	General Mills
Fiber One	0	General Mills
Nutri Grain (Almond Raisin)	7	Kellogg's
Nutri Grain (Wheat)	2	Kellogg's
Heartwise	9	Kellogg's
Kenmei Rice Bran	4	Kellogg's
Product 19	3	Kellogg's
Just Right	5	Kellogg's
Crispix	3	Kellogg's
All Bran	5	Kellogg's
Cracklin' Oat Bran	7	Kellogg's
Common Sense Oat Bran	10	Kellogg's
Oatbake (Raisin Nut)	8	Kellogg's
Oatbake (Honey Bran)	8	Kellogg's
Bran Flakes	5	Kellogg's
100% Natural Granola	7	Quaker
100% Granola (Raisin and Date)	8	Quaker

Reference

Friel, S. N., J. R. Mokros, and S. J. Russell. *Statistics: Middles, Means, and In-Betweens.* Palo Alto, California: Dale Seymour Publications, 1992; p. 87, Student Sheet 9.

INVESTIGATION 32
Students' Understanding of the Mean: A Minilecture

3The NCTM *Curriculum and Evaluation Standards for Teaching Mathematics* recommends that teachers involve students in the entire process of statistics: formulating questions, collecting data, and analyzing and interpreting data. By engaging students in a variety of problems, tasks, and activities that deal with statistics, teachers can provide opportunities for them to develop individual conceptions of the mean in addition to practicing the computational algorithm.

We typically teach the concept of the mean at the elementary-grades level as if our only curricular goal was for students to memorize and practice the algorithm. Instruction at the middle and high-school levels usually does little more with the concept. According to the Fourth Mathematics Assessment of the National Assessment of Educational Progress, or NAEP (Brown and Silver 1989), "most students in the seventh and eleventh grades appeared to understand technical statistical terms such as mean, median, mode, and range. However, there is evidence that they could compute the mean when asked for the average" (p. 28). The Second International Mathematics Study (Travers 1986) indicates that grade 8 mathematics teachers in the United States spend about four class periods teaching probability and statistics. At the grade 12 level, precalculus and calculus classes give little attention to the teaching of probability and statistics.

This minilecture is an overview of current thought on students' understanding of the mean in the middle and upper grades.

Transparency 32.a

Three main ideas need to be addressed in our discussion of statistics education. First, students must deal with statistics content in several forms: manipulation of objects, graphical representations, symbolic presentation of concepts, and real-world applications. Many of the problems we have tackled have approached statistics education at each of these levels. We manipulated objects as we collected data. We measured distances, counted raisins, and timed how long we could hold our breath. We represented data with pictures and graphs: line plots, bar graphs, and stem plots. We used an algorithm to calculate the mean of a data set. The graphs we created, as well as the statistical information from newspapers and magazines we gathered, provided good examples of real-world applications.

Materials

Transparencies 32.a through 32.h

The second idea we need to keep in mind is that students' backgrounds in statistics are most likely incomplete and probably contain inaccuracies. We can all think of times when students were unable to deal with statistical problems that were only slightly different from those in their textbook.

Third, the difficulties students encounter in learning statistics may persist into adulthood. Since little time is spent on statistical content at the middle- and high-school levels, we cannot assume that students will pick up what they need later on. One example of a concept that is difficult for many adults is that of the weighted mean.

Results of the Fourth Mathematics Assessment

The Fourth Mathematics Assessment included 16 items at grade 3, 23 items at grade 7, and 23 items at grade 11 that related to data organization, analysis, and interpretation. These items tended to be positioned toward the end of students' tests, so response rates were fairly low. However, the problems that students did attempt to answer can help us understand how they think about the mean.

Transparencies 32.b and 32.c

Student responses to problems 1 and 2 indicate that the concept of mean may not be as simple for students to understand as we might think. Participants may notice the difference in correct responses when the terms *mean* and *average* are used.

Transparencies 32.d and 32.e

Problems 3 and 4 involve weighted means and provide examples of misunderstandings that may persist into adulthood. Students often think that all given information is of equal importance.

Transparency 32.f

In problem 5, participants may recognize that the balance model of the mean helps to determine the appropriate score.

Transparency 32.g

Students' difficulty with problem 6 cannot be blamed entirely on their difficulty with the concept of the range. On two other problems, more than half the students correctly identified the range of a set of numbers.

Several conclusions can be drawn from the Fourth Mathematics Assessment data. First, students at all three grade levels performed poorly on items related to measures of center and variability. Second, many students in grades 7 and 11 do not understand terms such as *mean, median, mode,* and *range;* however, they can compute the mean when the term *average* is used. Third, most students had difficulty with problem-solving tasks and presentations of information that differed even slightly from those in their textbooks.

Transparency 32.h

Strauss and Bichler (1988) identified seven fundamental properties of the concept of the mean. The researchers created tasks to test children's understanding of each of these properties; 8-, 10-, 12-, and 14-year-old children solved the problems individually. Generally, 14-year-olds performed better than 12-year-olds, who performed better than 10-year-olds, who performed better than 8-year-olds.

1. The average is located between the extreme values. *Approximately half of the 8-year-olds and almost all of the 10-, 12- and 14-year-olds solved these problems correctly.*

2. The sum of the deviations from the average is zero. *Students at all age levels had difficulty with these tasks, and the 10- and 12-year-olds performed about the same.*

3. The average is influenced by values other than the average.

4. The average does not necessarily equal one of the values that was summed. *Almost all of the children solved these problems correctly.*

5. The average can be a fraction with no counterpart in physical reality. *About 40 percent of the 8-year-olds, 80 percent of the 10-year-olds, and almost all of the 12- and 14-year-olds solved these problems correctly.*

6. When one calculates the average, a value of zero, if it appears, must be taken into account. *This property was very difficult for the children, with only 25 percent of the 8-year-olds, 20 percent of the 10-year-olds, and 60 to 65 percent of the 12- and 14-year-olds correctly solving these problems.*

7. *The average value is representative of the values that were averaged.* *This property was also difficult for the children: very few of the 8-year-olds, 25 percent of the 10-year-olds, and 60 to 65 percent of the 12- and 14-year-olds solved these problems correctly.*

Summary

More research is needed on how students understand and recall the concept of the mean. We also need more classroom-based research on how certain teaching activities influence students' knowledge of the mean.

Discussion in the classroom is critical for developing students' mathematical ideas. Statistics is an ideal place in the curriculum for developing students' abilities to engage in discourse. Every student can contribute.

It is important that teachers understand that students often have difficulty with the concept of the mean and need many different types of problems spanning the various properties of the concept.

References

Brown, C. A., and E. A. Silver. "Data Organization and Interpretation." In *Results from the Fourth Mathematics Assessment of the National Assessment of Educational Progress*, edited by M. Lindquist. Reston, Virginia: National Council of Teachers of Mathematics, 1989; pp. 28–34.

Strauss, S., and E. Bichler. "The Development of Children's Concepts of Arithmetic Average." *Journal for Research in Mathematics Education,* 19 (1): 64–80 (1988).

Travers, K. J. (ed.). *Second International Mathematics Study: Detailed Report for the United States.* Champaign, Illinois: Stipes Publishing, 1986.

Three Ideas About Teaching and Learning Statistics

Students must deal with data in a variety of contexts, including

physical manipulation
pictorial and graphical representations
symbolic representations
real-world applications.

Students' prior knowledge of statistics may be incomplete and may be inaccurate.

Students' difficulties may persist into adulthood.

Problem 1

Inches of Snow in January	
Year	Inches of Snow
1970	15
1971	16
1972	17
1973	15
1974	15
1975	16
1976	16
1977	18
1978	15
1979	17
1980	15
1981	17
1982	16
1983	17
1984	15

a. What is the mode?
b. What is the median?
c. What is the mean?

Percent Correct (Response Rate)

	Grade 7	Grade 11
Question a	26 (.65)	40 (.41)
Question b	38 (.65)	47 (.41)
Question c	40 (.66)	41 (.72)

Reference

Brown, C. A., and E. A. Silver. "Data Organization and Interpretation." In *Results from the Fourth Mathematics Assessment of the National Assessment of Educational Progress*, edited by M. Lindquist. Reston, Virginia: National Council of Teachers of Mathematics, 1989; p. 29.

Problem 2

Here are the ages of six children:

13   10   8   5   3   3

What is the average age of these children?

Percent Correct (Response Rate)

Grade 7	Grade 11
46 (.94)	72 (.98)

Reference

Brown, C. A., and E. A. Silver. "Data Organization and Interpretation." In *Results from the Fourth Mathematics Assessment of the National Assessment of Educational Progress*, edited by M. Lindquist. Reston, Virginia: National Council of Teachers of Mathematics, 1989; p. 30.

Problem 3

Score	A = 4	B = 3	C = 2	D = 1	F = 0
Frequency	8	7	0	3	5

Which of the following procedures will give the average grade for the test scores given above?

a. $\dfrac{4 + 3 + 2 + 1 + 1}{18}$

b. $\dfrac{(8 \times 4) \times (3 \times 7) \times (2 \times 0) \times (1 \times 3) \times (0 \times 5)}{23}$

c. $\dfrac{(8 \times 4) + (7 \times 3) + (0 \times 2) + (3 \times 1) + (5 \times 0)}{23}$

d. $\dfrac{(4 \times 8) + (3 \times 7) + (3 \times 1)}{18}$

Percent Responding

Choice	Grade 11 (response rate = .84)
a	19
b	30
c	43
d	9

(This item was not included on the grade 7 test.)

Reference

Brown, C. A., and E. A. Silver. "Data Organization and Interpretation." In *Results from the Fourth Mathematics Assessment of the National Assessment of Educational Progress*, edited by M. Lindquist. Reston, Virginia: National Council of Teachers of Mathematics, 1989; p. 31.

Problem 4

Louise bought some packages of fudge:

2 pounds of vanilla for $0.90 per pound
3 pounds of chocolate for $1.60 per pound

What was the average cost per pound for this fudge?

Percent Responding

Answer	Grade 7 (response rate = .50)	Grade 11 (response rate = .58)
$0.50	17	24
$1.25	14	22
$1.32	12	20
$2.50	25	11
$3.30	6	3
$6.50	16	11
I don't know	14	9

Reference

Brown, C. A., and E. A. Silver. "Data Organization and Interpretation." In *Results from the Fourth Mathematics Assessment of the National Assessment of Educational Progress*, edited by M. Lindquist. Reston, Virginia: National Council of Teachers of Mathematics, 1989; p. 31.

Problem 5

Edith has an average (mean) score of 80 on five tests. What score does she need to get on the next test to raise her average to 81?

Percent Responding

Answer	Grade 11 (response rate = .48)
81	14
82	31
85	31
86	24

(This item was not included on the grade 7 test.)

Reference

Brown, C. A., and E. A. Silver. "Data Organization and Interpretation." In *Results from the Fourth Mathematics Assessment of the National Assessment of Educational Progress*, edited by M. Lindquist. Reston, Virginia: National Council of Teachers of Mathematics, 1989; p. 32.

Problem 6

Bill made the lowest score on the test. He only got 29 points. The teacher said the class mean was 65, and the range was 51. Jane made the highest score on the test. What score did Jane get?

Percent Responding

Answer	Grade 11 (response rate = .68)
51	12
65	29
80	43
94	16

(This item was not included on the grade 7 test.)

Reference

Brown, C. A., and E. A. Silver. "Data Organization and Interpretation." In *Results from the Fourth Mathematics Assessment of the National Assessment of Educational Progress*, edited by M. Lindquist. Reston, Virginia: National Council of Teachers of Mathematics, 1989; p. 33.

Seven Properties of the Mean

The average is located between the extreme values.

The sum of the deviations from the average is zero.

The average is influenced by values other than the average.

The average does not necessarily equal one of the values that was summed.

The average can be a fraction that has no counterpart in physical reality.

When one calculates the average, a value of zero, if it appears, must be taken into account.

The average value is representative of the values that were averaged.

Reference

Strauss, S., and E. Bichler. "The Development of Children's Concepts of Arithmetic Average." *Journal for Research in Mathematics Education*, 19 (1): 64–80 (1988).

INVESTIGATION 33
Classroom Assessment of Mathematics: A Minilecture

This minilecture is an overview of the state of mathematics assessment and the different forms assessment may take.

Transparency 33.a

Assessment has always been a part of classroom instruction. The demand for grades and accountability is part of the tradition of American schools; however, society has focused primarily on computational prowess, memorization of procedures, and problem solving designed to apply the discrete information or skills currently being taught. Testing—especially the high-stakes accountability measures administered by school systems and state agencies—has reinforced this narrow view. By reducing the complex issue of evaluating student achievement to a few numbers used to sort and classify students, traditional pencil-and-paper assessment has frequently influenced instruction in a negative manner.

Evidence comes from teachers themselves, who say they are not willing to teach what they feel would be most valuable or to use practices that would provide students with time for exploration and reflection because they feel pressured to provide practice for "the test." They have, in fact, aligned instruction with assessment expectations. They have provided drill and practice with testlike items and mirrored in daily classroom practices the no-talking, work-alone atmosphere of standardized tests.

Transparency 33.b

What is happening to cause such a renewed and intense interest in alternative assessment?

First, technology is changing the workplace and the skills needed by workers. The mathematics of previous decades is inadequate to prepare students for the modern workplace. Business and industry want workers who can make decisions based upon data, work cooperatively, and use mathematics to solve problems. Machines can do the computation, but workers must decide what needs to be done. The fastest-growing segments of our workforce—women and minorities—are those who traditionally have had the least background in the kinds of mathematics that are needed in the workplace of the twenty-first century.

Materials

Transparencies 33.a through 33.n

Teacher Notes

Every instructor has unique understandings and experiences to contribute. The text and transparencies are designed to support a minilesson on this topic; modify them as you wish.

Second, current practices do not seem to be working. Despite the need for mathematical understanding and expertise in so many aspects of society, it is still socially acceptable to "not be good in math." Many students take only the minimum required courses in high-school mathematics, and relatively few minority students choose to take higher-level courses. Colleges and universities report large numbers of students who need remedial mathematics courses before they are able to begin college-level work. Similarly, American students do not compare well on standardized measures with children from other nations, nor do they score well on national tests.

Third—and perhaps most important for change—there is growing support among those within the educational community, the national political arena, and local leadership for the reshaping of evaluation practices. The NCTM Standards are serving to provide a forum for discussion. While many of the recommendations are not new, they have given greater awareness to the complexity of mathematics learning, provided basic guidelines for evaluation, and lent a vision of what quality mathematics education needs to be. The Standards appropriately and emphatically support the classroom teacher as a decision maker and the primary evaluator of student learning.

Focus and Components of Assessment

If the momentum is to continue and prevail, change in assessment practices must be led by teachers who have developed expertise in a variety of assessment strategies and who are able to articulate the benefits. Teachers must know what their students have understood in the past in order to plan future lessons. Parents want to know how their children are progressing. School systems and state officials want to know how students compare across standardized measures and if the prescribed curriculum is being taught and learned.

Transparencies 33.c and 33.d

The primary focus of assessment is shifting away from sorting and classifying students to a much broader picture of what students know, understand, and can do. External evaluation, most frequently norm-referenced testing, is now viewed as just one of the components used to measure educational progress. We are moving from snapshots of student knowledge—pictures taken at one point in time and from only one angle—to motion pictures—examples gathered in context over time, and in many different ways.

We now look at assessment in terms of the many facets of mathematical achievement: concepts, factual information, skills, use of strategies, awareness of the importance and applications of mathematics, attitudes and inclinations, the ability to make connections and to communicate, and students' independent use of mathematics. In other words, evaluation is moving far beyond determining whether or not a student can "get the right answer." You may wish to illustrate each of the ideas on Transparency 33.d and invite participants to comment.

Teachers now strive for insight into students' methods of thinking and understanding in order to make instructional decisions as well as to describe the progress students are making. Alternative forms of assessment are being used more and more frequently as teachers discover how much and what kinds of information can be gathered from interviews and open-ended questions.

Transparency 33.e

There are many forms of alternative assessment. All require greater reflection on the part of the teacher in lending meaning to the student's response and thoughtful, professional judgment in assigning value and judging merit. All of the techniques appear to take more time than traditional methods do, or at least a reallocation of preparation and in-class time.

Alternative assessment techniques have similar features and, typically, four parts: the task or situation (question), the student's response, the teacher's interpretation and scoring of the response, and the way information about the response is recorded and results are reported.

It is not easy to identify and create rich, appropriate tasks, nor to design good questions. Because second and third questions are often determined by a student's response to the first query, interviews are hard to script. Assigning meaning to the student's response may also be difficult, and therefore some teachers hesitate to move to "How do you know?" or "Why do you think that?" questions. Because teachers are being asked to behave in ways they may never have seen modeled and to make assessment an ongoing process rather than a culminating event, the reallocation of their instructional time may feel uncomfortable. For these and other reasons, many teachers find their initial efforts to use alternative means of assessment frustrating.

You may wish to stop and discuss how the four components of assessment listed in Transparency 33.e fits with the activities and lessons teachers have been experiencing this summer. A good resource for a discussion on purposes and methods of assessment is the chart on pages 200–201 of the *Curriculum and Evaluation Standards for School Mathematics* (NCTM 1989).

Variations

There are commercial videotapes available in which interviews are modeled. Participants may wish to view some of these, then discuss "good questions" and sample interview record forms. *Mathematics Assessment: Myths, Models, Good Questions, and Practical Suggestions* (Stenmark 1991) is a reader-friendly resource that could help to expand the discussion of assessment beyond a brief lecture.

The movement for alternative assessment has a broad base of leadership. Teachers, researchers, and administrators are helping to work through the traditional testing issues of grades, validity, and reliability. They are designing tasks and assigning meaning to student responses. They are developing ways to score and report their findings.

Types of Alternative Assessment

Many important forms of alternative assessment are being widely discussed; they are not mutually exclusive and are frequently combined. Let's briefly look at them, beginning with a few notes on traditional forms of assessment.

Transparencies 33.f and 33.g

Traditional Pencil-and-Paper Evaluations

- greater focus on criterion-referenced tests

- more emphasis on higher-order thinking

- computation evaluated within a context rather than with symbols alone

- fewer multiple choice and more free response

Informal and Formal Interviews

- may be scheduled as quiet one-on-one interviews or within the routine of class activities

- usually short (3–10 minutes) with a specific purpose

- likely to be the best method of assessing students' thinking and understanding, especially for younger students

- allows teachers to explore attitudes, divergent thinking, misconceptions, and unique solutions

- teachers need questions in mind but should be ready to follow student leads

- should be nonjudgmental (a time for gathering information, not for correcting mistakes or instructing)

- important to keep some type of dated record

Classroom Observations

- informal or planned for a specific concept or skill

- notations might occur during group work, single recitation, or within other class structures

- a student's disposition, ability to communicate, and unstructured use of (making connections and applying) mathematics may be observed

- record keeping takes many forms

- observations frequently identify children for one-on-one or small-group interviews

- important to describe what is observed without lending interpretation to every annotation

Performance Tasks

- usually similar to instructional tasks

- looks at process as well as product

- may be short activities or extended investigations

- looks at use of mathematics rather than discrete skills

- may be evaluated holistically or analytically

- links academic and real-world mathematics

- may be individual or group

- usually incorporates open-ended questions

Open-ended Questions

- opportunities to examine students' thinking; divergent thinking encouraged

- both routine and nonroutine questions

- allows focus on process as well as solution

- insight into students' understanding

- usually scored with a rubric—holistically or analytically

Teacher Notes

A discussion of rubric scoring may be inserted here or you may wish to spend more time with it later. It is important to stress that a general rubric is just that—a generic guideline used to facilitate consistency in scoring. A rubric is usually written for each open-ended question or task, spelling out exactly the expectations for each point of the students' scores.

Journals and Logs

Checklists and Computer Programs

- used to track student performance

Student Self-assessment and Conferencing

Transparencies 33.h, 33.i, 33.j, 33.k, and 33.l

Portfolios

Transparencies 33.m and 33.n

References

Curriculum and Evaluation Standards for School Mathematics. Reston, Virginia: National Council of Teachers of Mathematics, 1989.

Assessment in the Mathematics Classroom, eds. N. Webb and A. Coxford. Reston, Virginia: National Council of Teachers of Mathematics, 1993.

Mathematics Assessment: Myths, Models, Good Questions, and Practical Suggestions, ed. J. Stenmark. Reston, Virginia: National Council of Teachers of Mathematics, 1991.

Assessment Standards for School Mathematics. Reston, Virginia: National Council of Teachers of Mathematics, 1995.

Report Card for Julie Kate

Communication Skills	C+
Mathematics	B
Science	B–
Social Studies	A–

Julie is a cooperative student. She needs to work on remembering to bring in her homework.

Why Is Assessment Changing?

Technology

Skills needed by workers

Composition of workforce and society

Poor test scores and comparisons

Few students taking advanced math

Students unprepared for work and college

Political awareness

Recognition of test limitations

NCTM Standards

Teacher leadership

Focus of Assessment

Traditional:

sort and classify students

Current direction:

teachers gathering information in order to better serve students

Assessing Mathematics

Facts

Concepts

Skills

Connections

Strategies

Awareness

Attitudes and inclinations

Communication

Independence

Four Components of Assessment

Task or situation (question)

Student's response

Teacher's interpretation of response

Recording and reporting

Methods of Assessment

Traditional pencil-and-paper evaluations

Informal and formal interviews

Classroom observations

Performance tasks

Open-ended questions

Journals and logs

Checklists and computer programs

Self-assessment and conferencing

Portfolios

Record-keeping Systems

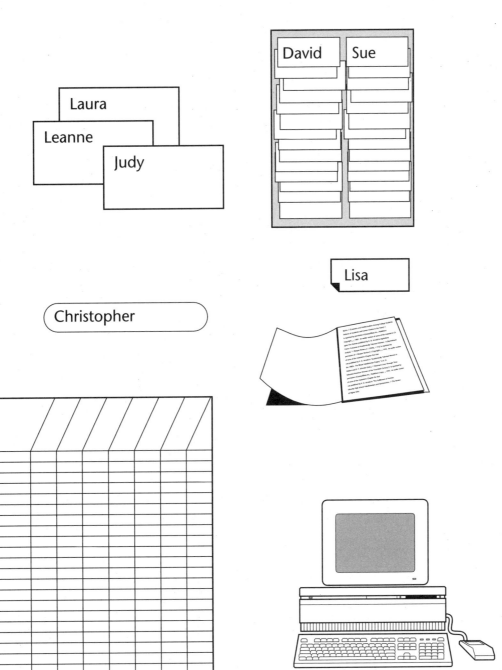

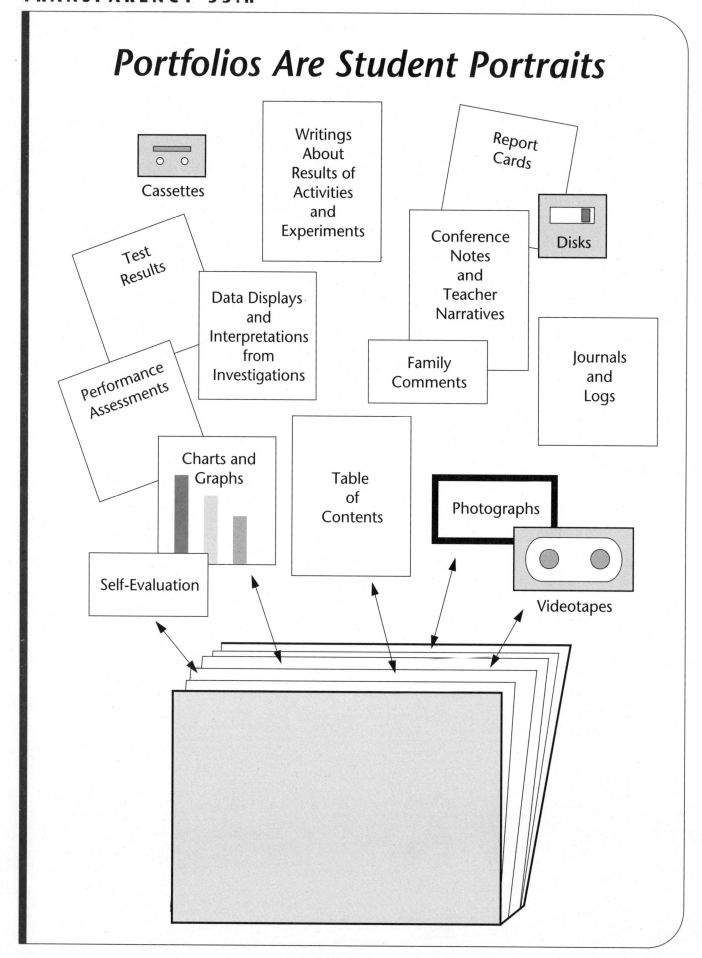

Showcase Component

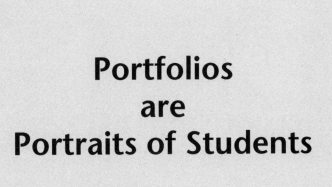

Portfolios are Portraits of Students

Benchmark Papers

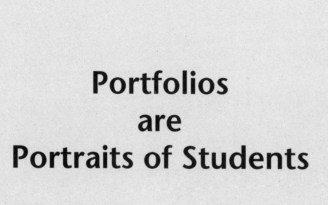

Commentaries and Anecdotal Records

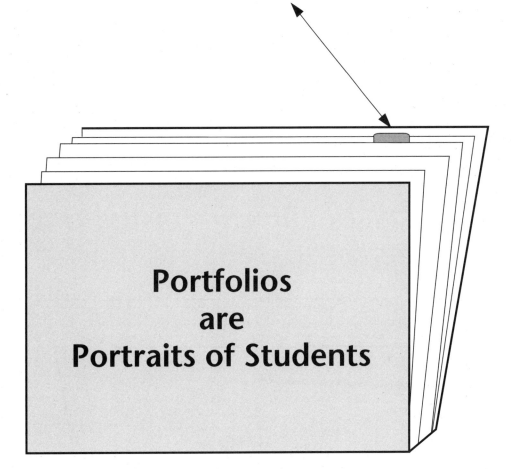

Working Portfolio

1. *Save samples throughout the grading period.*

2. *Involve the student in choosing work for the portfolio.*

3. *Make final selections that present a broad picture.*

4. *Consider options in evaluating the material.*

Tips for Assessment

Make ongoing assessment an integral part of your teaching.

Involve students in self-evaluation.

Develop problem situations that require application of several mathematical ideas. Consider some "group tests."

Include an open-ended question on all pencil-and-paper tests.

Assess mathematical disposition, different kinds of mathematical thinking, as well as student interest, curiosity, and inventiveness.

Incorporate technology into assessments through audiotapes, videotapes, and computers.

Become more reflective in your evaluations.

How Do We Begin?

Decide what is most important.

Begin small.

Fit into your routine.

Develop comfortable record-keeping methods.

Stick with the task.

Revise and personalize.

Find a buddy.

INVESTIGATION 34
Linking Probability and Statistics: A Minilecture

This minilecture is an introduction to probability.

Probability holds an important place in the field of statistics. While probability can be studied as a mathematical discipline in its own right—with its own foundations, definitions, and important results—the practitioner of statistics tends to use probability in two principal ways.

- Probability is a tool for understanding sampling issues in statistics and influences the design of experiments and surveys.

- Probability helps us to interpret inferences that researchers make from data gathered through experiments and surveys.

An intuitive way to think about probability is as *the likelihood that something will happen*. An impossible event has a probability of 0, while a certain event has a probability of 1. Other events have other probabilities, all between 0 and 1, and their likelihood of occurrence is small or large depending upon whether their probability is "close to 0" or "close to 1." A relative probability of 1 is perfect probability, or 100 percent probable.

One way to interpret probability is to use the concept of relative frequency to help analyze the likelihood that one or more outcomes from an event will occur. When we compute relative frequency, we are describing a *model* of the variability, or randomness, of an event and the possible outcomes associated with that event. When we toss a coin, roll a pair of dice, or spin a spinner, we observe different outcomes for each event. If we gather data about the occurrence of each outcome over a number of trials, we can apply the concept of relative frequency to make good estimates about how often each outcome is likely to occur.

Let's define *relative frequency* and see how it fits with our use of probability by looking back at data we gathered while doing Investigation 22, *Raisins Revisited*.

Materials

Transparencies 34.a, 34.b

Sample Questions

These questions can be asked followed by a discussion of what "chance" means in each situation.

There is a 40 percent chance of rain today. What are the chances of rain tomorrow?

A basketball player makes 70% of her free throws. What are the chances that she will make all five free throws in a game?

What are the chances that a particular candidate will win a particular election?

What are the chances that I will win the lottery?

Transparency 34.a

From the frequency chart, we can see that 8 boxes of raisins have 30 raisins each. We say that the *frequency of boxes with 30 raisins in them is 8*. The frequency of boxes with 26 raisins is 1, and the frequency of boxes with 20 raisins is 0.

Relative frequency can be expressed as a fraction, a decimal, or a percent. The corresponding relative frequency of boxes with 30 raisins is computed as 8 in 29, $8/29$ or 28 percent; the relative frequency of boxes with 26 raisins is $1/29$ or 3 percent; and the relative frequency of boxes with 20 raisins is $0/29$ or 0 percent. Notice that the frequency is simply another word for *count*, while relative frequency can be thought of as the *fraction or percentage of the total count*, since there is a total of 29 boxes in this particular sample of raisins.

In this sample, the relative frequency of boxes with 30 raisins is eight times *greater than* the relative frequency of boxes with 26 raisins, and thus would be much *more likely to occur* (that is, has a higher probability of occurring). Be aware, however, that a sample of boxes of another brand of raisins might not have the same distribution of relative frequencies. You can compare the results of your *Raisins Revisited* activity to the data plotted here to verify this.

Recall that when we performed the *Raisins Revisited* activity, we asked this question: If we were to open another half-ounce box of raisins, how many raisins would you predict will be in that new box? A good answer would certainly be one that used the existing raisin data as a guide. In our example, we may predict 30 or 31 raisins, because the highest relative frequencies are for boxes with 30 and 31 raisins. In fact, the event "there are 30 or 31 raisins in the box" has a relative frequency of $15/29$, which is greater than 50 percent.

Transparency 34.b

We can look back at data from other activities we have done and use the differences we observed in the relative frequencies to model the likelihood that events will occur. The plot here shows the distribution of the color of cats' eyes from a sample of 48 cats. Use the plot to predict the eye color of the next cat we see. Assume that the next cat is sampled in the same way the original cats were sampled.

Compute the relative frequency of each eye color by dividing the frequency of the color by 48. Then, compare the relative frequencies. Which eye color do you predict will be the *most likely* to occur for the next cat we sample? Surely not gold or multicolored, since their relative frequencies are each $1/48$ or 2 percent.

Summary

Relative frequencies found in simple dice and coin experiments and games will be explored through the statistical framework introduced in our earlier investigations, namely *pose the question, collect the data, analyze the data,* and *interpret the results.*

The concept of probability will be introduced through a series of statistical investigations that engage participants in various experiments. Data gathered from the outcomes of each experiment may be analyzed by sketching appropriate graphs and computing the relative frequencies of the data over a large number of repetitions of the experiment in order to consider likelihoods of occurrence of outcomes.

The explorations associated with these experiments will be tied to an introduction of the concept of randomness. Finally, we will briefly consider how the use of random-data generators such as dice can help us to understand the concept of sampling and its use in statistics as a way to collect data.

Raisin Data: Frequency

Raisins in Half-Ounce Boxes of Brand X Raisins

```
                              X
                              X    X
                              X    X
                    X         X    X
                    X         X    X
          X    X    X    X    X
          X    X    X    X    X    X
     X    X    X    X    X    X    X
    ―――――――――――――――――――――――――――――――
     26   27   28   29   30   31   32
```

Number of raisins

Number of Raisins in Box	Frequency
26	1
27	3
28	5
29	3
30	8
31	7
32	2

Total = **29**

Number of Raisins in Box	Relative Frequency		
26	$1/29$	0.03	3%
27	$3/29$	0.10	10%
28			
29			
30			
31			
32			

Cat Data: Frequency

Eye Color in a Sample of Cats

```
                                            ×
                                            ×
                                            ×
                                            ×
                                            ×
                              ×             ×
                              ×             ×
                              ×             ×
                              ×             ×
                              ×             ×
                              ×             ×
                              ×             ×
                              ×             ×
                              ×             ×
                ×             ×             ×
                ×             ×             ×
                ×             ×             ×             ×
                ×             ×             ×             ×
  ×      ×      ×             ×             ×             ×
─────────────────────────────────────────────────────────────
 Gold Multicolored Green-gold Yellow      Green         Other
                          Eye Color
```

Eye Color	Relative Frequency		
Gold	$1/48$	0.02	2%
Multicolored	$1/48$	0.02	2%
Green-gold	$6/48$	0.13	13%
Yellow	$16/48$	0.33	33%
Green	$21/48$	0.44	44%
Other	$3/48$	0.06	6%

INVESTIGATION 35

True-False Test

Overview

Participants will probably be familiar with true-false questions as used on tests. Many may believe that a score of 7 or 8 correct on a test of ten true-false questions indicates "good understanding" of the material. In reality, a score of 7 or 8 could easily be achieved by a person who knows only half of the answers and then guesses on the remaining five questions.

This investigation illustrates this situation and provides an opportunity to explore how probability affects an evaluation practice that may be used routinely.

Assumptions

Participants have little background in the study of probability. They have some experience with the process of statistical investigation.

Goals

Participants explore probability as it relates to true-false tests. In particular, they

- explore the concept of randomness

- use the concept of relative frequency to interpret the probability of an event; that is, the likelihood of occurrence of outcomes for an event

Developing the Activity

You will have to prepare the ten-item true-false test for participants in advance—a fun activity in itself. Make the first five questions easy enough for everyone to answer correctly (for example, There are 50 states in the United States; Exactly 27 people are in this room). Choose the last five questions so that only you know the answers (for example, Total enrollment in elementary and secondary schools in North Carolina in fall 1990 was 1,086,888; In the presidential election of 1828, Andrew Jackson defeated John Quincy Adams

Materials

A 10-item true-false test (prepare in advance)

Teacher Notes

This activity could lead to a fruitful discussion of multiple-choice testing and what probability tells us about results that should or should not be surprising in such a test format.

by a popular-vote margin of 138–134); carefully research the questions in a world almanac or similar reference book.

Pose the Question

Are true-false tests reliable measures of knowledge?

Give participants a few minutes to discuss the question, then indicate that they will now take a true-false test consisting of ten questions.

How many questions would you expect to answer correctly if you were certain of the answers to the first five questions but had to guess for the remaining five?

Have participants record their guesses and discuss their thinking. Discussion of this question might center around the notion of a "passing grade" on such a test (such as 7, 8, 9, or 10 correct answers).

Collect the Data

Have participants number a paper from 1 to 10. Read aloud the true-false test you have created, having participants record their answers as *T* or *F* to each question.

Have participants score their own tests as you read the correct answers. If the group consists of only a few participants, it might be necessary for each participant to take the test two or three times (independently and randomly choosing a string of *T* and *F* answers to the last five questions) to generate enough data to show a pattern in the outcome of the event "passing the test."

Analyze the Data

With participants, construct a line plot to show the number of correct answers.

How many people got a score of 7 or more?

Record this result as a percent of the total number of participants. Discuss how many participants "passed the test" and why the results were what they were.

Is it possible to describe the expected *outcomes and relate that description to the* actual *outcomes?*

 Variations

Another way to carry out this activity that has met with great success is to announce the ten-question test and then have the questions read in another language, such as Arabic or Chinese, that almost certainly none of the participants understands. This puts everyone on the same level, and they must truly guess at the answers. Give participants the answers to the first five questions and have them guess at the remaining five. This models the situation in which the test-taker knows half the answers.

If we know the first five responses are always correct, then we need to think about what the likelihood of occurrence is of getting 0, 1, 2, 3, 4, or 5 correct responses in the second five questions when guessing is used as the response method.

With participants, develop a table of the possible outcomes. For example, there is only one way for the number of correct responses to be 0 (when all five questions are answered incorrectly); five ways for the number of correct responses to be 1 (when any one of the five questions is answered correctly); and ten ways for the number of correct responses to be 2 (when any combination of a pair of questions is answered correctly).

Number of Correct Responses	Number of Ways for This to Occur
0	1
1	5
2	10
3	10
4	5
5	1

Possible outcomes = **32**

If "7 correct" constitutes passing, then there are 10 + 10 + 5 + 1 = 26 out of 32 ways to pass the test. So even if the test-taker knows the answers to only 50 percent of the questions, there is an 81 percent chance he or she will pass the test.

Interpret the Results

Encourage participants to discuss the value of the true-false test as a measure of knowledge.

INVESTIGATION 36
Removing Markers from a Number Line

Overview

Participants play a game using a pair of dice. By playing the game many times, they discover that there are differences in likelihood of the outcomes of rolling a pair of dice, and they use these principles to develop a strategy for winning the game.

Assumptions

Participants will be familiar with the possible outcomes when a pair of dice is rolled (the sum of the dots on the upward-facing side of each die is the outcome). They are familiar with a standard number line.

Goals

Participants explore the concept of probability. In particular, they

- use observational skills

- develop a strategy for improving their chances of winning the game

- gain insight into the fact that, when rolling a pair of dice, not all events are equally likely

- formulate an intuitive notion of the probability that an event will occur when tossing dice

Reference

Burns, M., and B. Tank. *A Collection of Math Lessons for Grades 3 Through 6.* New Rochelle, New York: The Math Solution Publications, 1987.

Developing the Activity

Participants will play a game intended to increase their experience with the concept of probability.

Materials

Markers (chips or cubes, 11 per person), pair of dice (1 per pair of players), colored markers

Handout 36.1

Pose the Question

Describe the number line, the dice, and how outcomes will be recorded (it is likely that at least one person has had little or no experience with dice). Distribute Handout 36.1 and 11 markers to each participant and a pair of dice to each pair.

This is a game of strategy for pairs of players. Let me describe the rules.

Each of us begins with 11 markers and a line numbered 2 through 12. To begin, place your markers on your number line in any order you wish (for example, place one chip at each place, or place multiple chips on some numbers and single chips on others).

Each player tosses the dice; the one rolling the highest sum goes first. The first player rolls the dice and records the sum of the dots on the upward-facing sides of each die. If he or she has a marker on the number line at that integer, the player removes the marker. Play passes to the second player. The game continues, with the dice being rolled and markers being removed (if the sum on the dice matches a marker's position) and play alternating between players. The winner is the first person to remove all of his or her markers.

Let's play one round of the game together. Place your 11 markers on your number line. I'll toss the dice and read the sums. We will play until we have one or more winners.

Collect and Analyze the Data

Have participants get into pairs and play the game. The first player to remove all the markers from his or her number line wins.

Keep track of each sum you roll at the bottom of your game sheet. We will use your collection of sums in the next investigation.

The more that participants play the game, the more they will refine their strategy for winning. For example, placing one marker on each of the numbers 2 through 12 might be an early strategy, but over time will probably be found to be less effective and thus abandoned.

Interpret the Results

Have participants describe winning strategies and their reasoning.

The next investigation, *What Are the Odds?* will formalize participants' understanding of probability as it pertains to dice.

Teacher Notes

Certain strategies for placement will clearly lead to winning the game quickly, while others will not.

Removing Markers Game Sheet

2	3	4	5	6	7	8	9	10	11	12

Sums Rolled

1. 16. 31. 46. 61. 76.
2. 17. 32. 47. 62. 77.
3. 18. 33. 48. 63. 78.
4. 19. 34. 49. 64. 79.
5. 20. 35. 50. 65. 80.
6. 21. 36. 51. 66. 81.
7. 22. 37. 52. 67. 82.
8. 23. 38. 53. 68. 83.
9. 24. 39. 54. 69. 84.
10. 25. 40. 55. 70. 85.
11. 26. 41. 56. 71. 86.
12. 27. 42. 57. 72. 87.
13. 28. 43. 58. 73. 88.
14. 29. 44. 59. 74. 89.
15. 30. 45. 60. 75. 90.

INVESTIGATION 37

What Are the Odds?

Overview

This activity is designed to illustrate the frequency of occurrence of each possible sum (2, 3, 4, 5, 6, 7, 8, 9, 10, 11, and 12) on rolls of a pair of dice.

Assumptions

Participants are familiar with recording, on a number line, the frequency of sums on a roll of a pair of dice.

Goals

Participants explore the concept of probability. In particular, they

- gain experience in collecting data
- develop skills in displaying data
- develop skills in making conjectures about data

Developing the Activity

Participants will use data they generate to further explore the concept of frequency.

Pose the Question

What does the relative frequency of sums on dice tell us about winning strategies in games?

Collect the Data

Participants work in teams of two, with one person rolling the dice and the other recording the outcomes on a line plot numbered 2 through 12. After two minutes, the participants trade roles and continue for two more minutes.

Materials

Pairs of dice, *Removing Markers Game Sheet*, blank transparency film, calculators

Handout 37.1

Variations

This activity can also be done with interlocking cubes. Each team makes cube towers resting on a number line, creating a three-dimensional line plot. Teams compute relative frequencies for later reference. Teams combine all their 2-towers into a single 2-tower, all their 3-towers into a single 3-tower, and so on. They can then place these towers side by side on the floor.

Line plots will look something like this:

```
                                            ×
                            ×       ×       ×
  ×           ×     ×   ×   ×       ×
  ─────────────────────────────────────────────
  2   3   4   5   6   7   8   9   10  11  12
```

Analyze the Data

As a group, compare the results gathered by the teams looking for similarities and differences.

After discussion about how the teams' results are similar to or different from each other, combine the results on a single number line on a transparency. Compute the relative frequencies of each outcome by dividing the number of times the given sum occurred by the total number of times the dice were rolled.

Interpret the Results

Make connections with the game played in Investigation 36, *Removing Markers from a Number Line*. Discuss the differences in the relative frequencies. For example, a sum of 2 (with a relative frequency of $\frac{1}{36}$) is much less likely to occur than a sum of 6 (with a relative frequency of $\frac{6}{36}$). This helps you decide where to place your markers. Discuss the fact that the sum of all the relative frequencies is 1, or 100 percent.

The theoretical distribution of outcomes when a pair of dice is rolled (assuming the dice are fair) is shown on Handout 37.1. A 6×6 matrix of the 36 possible outcomes when a pair of dice is rolled represents the collection of all possible outcomes. Assuming all of these are equally likely, the theoretical distribution is easily computed.

How does your newly computed relative frequency distribution compare to the theoretical distribution shown on the handout?

HANDOUT 37.1

Rolls of a Pair of Dice

Possible Outcomes

		Die 1					
		1	2	3	4	5	6
Die 2	1	2	3	4	5	6	7
	2	3	4	5	6	7	8
	3	4	5	6	7	8	9
	4	5	6	7	8	9	10
	5	6	7	8	9	10	11
	6	7	8	9	10	11	12

Theoretical Distribution of Outcomes

					X					
				X	X	X				
			X	X	X	X	X			
		X	X	X	X	X	X	X		
	X	X	X	X	X	X	X	X	X	
X	X	X	X	X	X	X	X	X	X	X
2	3	4	5	6	7	8	9	10	11	12

Sums

$1/36$ $2/36$ $3/36$ $4/36$ $5/36$ $6/36$ $5/36$ $4/36$ $3/36$ $2/26$ $1/36$

Relative frequencies

INVESTIGATION 38

Fair Games

Overview

The two games in this activity are played with a standard pair of dice. In the first game, the *sum* of the dots on the dice is the score; in the second game, the *product* of the dots on the dice is the score.

Assumptions

Participants can accurately compute sums and products of dots appearing on a pair of dice and can keep accurate records of wins and losses. They have had some experience with thinking about the likelihood of occurrences of sums on a pair of dice.

Goals

Participants explore the concept of probability. In particular, they

- gain experience with the concept of randomness

- begin to understand the notion of fair versus unfair games

- gain reinforcement using the concept of relative frequency

Reference

Bright, G. W., J. G. Harvey, and M. M. Wheeler. "Fair Games, Unfair Games." In *Teaching Statistics and Probability*, eds. A. P. Shulte and J. R. Smart. Reston, Virginia: National Council of Teachers of Mathematics, 1981.

Developing the Activity

The activity begins with a discussion of the notion of a "fair game" and what is meant by the phrase. Participants then play two dice games: Sum of Dots and Product of Dots. Each team rolls a pair of dice 20 times and records the outcomes. The 20 rolls make up the data sample.

Materials

Pairs of dice, calculators

Handout 38.1

Pose the Question

Begin with a discussion of the concept of a "fair game."

What is meant by a "fair game"?

Dialogue might initially center on the simplest of all fair games—tossing a fair coin—and how *fair* might be defined in this example.

Collect the Data

The games are played by gathering *sums* of dice for the first game and *products* of dice for the second game. Explain the rules, then let participants play both games.

In the Sum of Dots game, one player in each team of two chooses even sums and the other chooses odd sums. Each pair rolls the dice 20 times and records the results on Handout 38.1. The winner is determined by whether even or odd sums appear most often.

In the Product of Dots game, one player in each team of two chooses even products and the other chooses odd products. Each pair rolls the dice 20 time and records the results. The winner is determined by whether even or odd products appear most often.

Analyze the Data

Have teams calculate the percentages of even and odd sums, and then of even and odd products, that they obtained in their games.

Do you think these games are fair?

Interpret the Results

As a class, combine the results of all the teams and compute percentages of wins for even sums and products and then for odd sums and products.

Now what do you think about the fairness of these games?

Summary

When a pair of dice is rolled, there are 6 possible even sums (2, 4, 6, 8, 10, and 12) and only 5 possible odd sums (3, 5, 7, 9, and 11). A naive dice-roller might assume that the Sum of Dots game is unfair.

This is a common assumption, since the sum of dots for a single roll is *more likely* to be even than odd. However, what is important is the sum of the relative frequencies for odd outcomes and even outcomes.

When the combined results are displayed, the fairness of Sum of Dots and the unfairness of Product of Dots should be evident.

In the Product of Dots game, there are many more than 11 possible outcomes. If there is interest, a discussion of the theoretical distribution of products could ensue by constructing a chart similar to the one for sums shown on Handout 37.1.

HANDOUT 38.1

Fair Games Data Sheet

Roll	Outcome	Sum	Even	Odd	Roll	Outcome	Product	Even	Odd
	Example: (2, 5)	7		✓		(6, 4)	24	✓	
1									
2									
3									
4									
5									
6									
7									
8									
9									
10									
11									
12									
13									
14									
15									
16									
17									
18									
19									
20									

Total number of even sums =   Total number of even products =

Total number of odd sums =   Total number of odd products =

Relative frequency of even sums =   Relative frequency of even products =

INVESTIGATION 39

What's in the Bag?

Overview

Participants are introduced to sampling in this activity. The question they investigate—What is the percentage of x-colored chips in the bag?—nicely models questions that political and social scientists ask every day, such as:

- What proportion of the population approves of x?

- What proportion of the voters would vote for x if the election were held today?

This activity is an important introduction to the concepts of sampling and estimating population measures.

Assumptions

Participants are familiar with taking samples without bias and recording results accurately.

Goals

Participants explore the concept of sampling. In particular, they

- gain experience in sampling
- play a game to obtain information about a population

Developing the Activity

A large number of colored chips, in two colors, will represent the population under study. Participants will be asked to look at a sample drawn from this population to determine the distribution of the two colors, without sorting and counting the entire population. By carefully combining the results in different ways, this activity introduces ideas about variability and its relationship to sample size.

Materials

Paper bags containing a large number of colored chips of the same size and weight (in 2 colors, 1 bag per pair of participants), chart paper, stick-on notes or stick-on dots (in 2 colors), blank transparency film, calculators

What's in the Bag? **343**

Prepare the paper bags of chips in two colors—let's assume red and blue—in advance. Fix the percentage of the colors in every bag at the same "nice" fraction (not too close to 0 or 1); 40 percent and 60 percent works well. You will need one bag of chips for each pair of participants, all with the same distribution—that is, for example, a bag with 40 reds and 60 blues, 60 reds and 90 blues, or 80 reds and 120 blues.

Pose the Question

Present a paper bag of chips to the group. Before beginning this exercise, be certain that the same number of chips with the same color distribution are in each bag.

This paper bag contains chips of two colors, red and blue. Your job is to predict the distribution of those colors—that is, to predict what percentage of the chips in the bag is red and what percentage is blue.

You will not be able to look at all the chips in the bag. In fact, you will draw only one chip at a time, note its color, and replace it; then you will shake the bag to mix the chips again. You will continue this process until you have sampled and recorded the data from 50 chips. So that we will be able to group your "draws" later, please keep track of the order in which you draw the colors. You may want to number a list from 1 to 50 and write the color drawn next to each.

Collect the Data

Provide each team of two participants with a bag of chips.

Each person draws one chip at a time from the bag, notes its color, replaces it, and then shakes the bag to mix the chips again. Each person samples and records data from 50 draws, keeping track of the order in which the colors were drawn. (If the number of participants is small, have them each complete 100 draws so that enough of a sample is drawn for patterns to emerge when the data are analyzed.)

Analyze the Data

Given that you drew only one chip at a time, how is it possible for you to make your prediction?

Point out that the individual drawings of chips are all independent of each other in the same way that individual tosses of a coin or individual rolls of a pair of dice are independent events.

Because the draws were independent, we can pool all our draws and then consider samples of various sizes—just as we combined the results of dice rolls in earlier activities.

Pooling can be done in sample sizes of 5, 10, and 25 in the following way: assuming each participant has recorded the color data from 50 chips, the sample could be represented as 10 samples of size 5 (the first 5 draws are sample 1, the second 5 draws are sample 2, and so on); 5 samples of size 10 (the first 10 draws are sample 1, the second 10 draws are sample 2, and so on); and 2 samples of size 25.

Have each participant analyze his or her data in each of the following ways:

- **Case A: Sample sizes of 5.** Compute the percentage of red chips pulled for draws 1–5, then compute the percentage of red chips pulled for draws 6–10, and continue until the percentage of red chips has been computed for each of the 10 samples of 5 chips. The possible percentages for each sample are 0, 20, 40, 60, 80, and 100 percent. As a group, record the data on a line plot.

- **Case B: Sample sizes of 10.** Compute the percentage of red chips pulled for draws 1–10, then compute the percentage of red chips pulled for draws 11–20, and continue until the percentage of red chips has been computed for each of the 5 samples of 10 chips. The possible percentages for each sample are 0, 10, 20, . . . , 80, 90, and 100 percent. Record these data on the line plot.

- **Case C: Sample sizes of 25.** Compute the percentage of red chips pulled for draws 1–25, then compute the percentage of red chips pulled for draws 26–50. The possible percentages for each sample are 0, 4, 6, . . . , 92, 96, and 100 percent. Record these data on the line plot.

When the percentage of red chips has been computed for each sample size and the percentages for the whole group have been graphed, the notion of sampling distribution should become clear. Because each participant was sampling randomly and independently, the percentages of red chips they obtained will vary—not only from participant to participant, but also by the size of the sample. In Case A, the only possible sample percentages of red chips are 0, 10, 20, . . . , 80, 90, and 100 percent; however, in Case C, the possible sample percentages of red chips are 0, 4, 6, . . . , 92, 96, and 100 percent.

You may want to construct, on a transparency, a separate graph of the sample percentages found by the participants for each sample size. Alternatively, you could have participants construct three line plots—one for each sampling size—on chart paper using stick-on dots or notes in two colors. They could mark their sample percentages as the class reports its data, building the line plots

Variations

You may want to collect data from 50 draws for each participant and then explore the distribution of sample composition for different sample sizes one at a time. You would begin with Case A, making the display, discussing the distribution, and computing means, medians, and so on. Then you would consider Case B, completing the computation of percentages, displaying, and reflecting on how Case B compares with Case A. This discussion would continue as you moved to Case C.

Variations

M&M's candies have a planned percentage distribution *(don't tell participants this until after the investigation!)*:

Red	20%
Brown	30%
Tan	20%
Yellow	10%
Green	10%
Orange	10%

Get a large bag of M&M's candies (16 ounces or larger), and have each participant draw a sample size of 20, record the distribution of colors in the sample, then return the candies to the bag. Combine the outcomes for each participant (this assumes at least 24 participants). Can participants predict the distribution of colors from their sampling?

A variation of this activity is to give a single-serving bag of M&M's candies to each participant (this is each person's sample); have them determine the number of each color and combine outcomes; and, again, ask whether they can predict the distribution of colors from their samples.

in the process. The notion that variability decreases as sample size increases can be seen by graphing the distributions of different sample sizes on overhead transparencies. If the scales of the horizontal axes are the same, the graphs can be superimposed so that the relationship of sample size to variability can be observed. Participants certainly will not have obtained identical percentages of red chips in their samples, but there will be a clear pattern in the whole group's distribution for each sample size investigated.

Interpret the Results

Encourage participants at each stage, for each sample size, to estimate the percentage of red chips in the bag. They will begin to see that the pattern of their sample percentages "tightens" around the true percentages of red chips in the bag as graphs of larger samples are observed. Said another way, they can have more confidence in their estimates with larger samples.

Much of the data-analysis theory that participants have encountered in previous activities can be brought to bear on the combined data sets of sample percentages collected in this activity. Graphical representations—such as line plots and histograms—can be used to illustrate the distribution of the sample percentages of red chips. Numerical measures—such as mean and median—can be computed from the distribution and used to estimate the true (unknown) percentage of red chips in each bag.

INVESTIGATION 40

Choosing Samples

Overview

This activity builds on participants' experience with sampling distributions in Investigation 39, *What's in the Bag?* and demonstrates how the model of randomness illustrated in rolling pairs of dice can be used to simulate drawing a colored chip from a bag as they work to obtain a random sample. The notions of sampling both with and without replacement will be considered, although participants' work will be done with replacement to mirror the *What's in the Bag?* investigation. In addition, participants will have an opportunity to work both with random samples and convenience samples in order to understand the differences between these two kinds of sampling.

Materials

Pairs of 10-sided dice (in 2 colors; 1 per pair), calculators

Handouts 40.1 through 40.4

Assumptions

Participants are familiar with taking samples without bias and can record results accurately.

Goals

Participants explore the concept of sampling. In particular, they

- gain experience in sampling
- understand the distinction between random sampling and convenience sampling

Reference

Moore, D. S. *Statistics: Concepts and Controversies*, 3rd ed., New York: W. H. Freeman, 1990.

Developing the Activity

In the database of 100 cats (see Handout 40.1), the cats are numbered from 1 to 100. For purposes of drawing a random sample, we need to think of them as numbered 00 (which maps to 01 on the handout) through 99 (which maps to 100 on the handout). Summary statistics for the entire database are provided (see Handout 40.2) so that participants will be able to compare their sampling results with those of the population. As part of the investigation, participants will create at least one sampling distribution for several samples taken from the database *with replacement*.

Pose the Question

Distribute Handout 40.1.

Here is a database of information about 100 cats. We want to look at a sample of data from these cats and compare it to the population.

One way to do this is through convenience sampling. *If we were to look at a convenience sample, we might look at all the females, all the males, just at kittens (8 months or younger), or just at cats with green eyes. Would these be representative samples for the population of cats as a whole? Explain your reasoning.*

You might want to share the data from each of these convenience samples (which are displayed on Handouts 40.3 and 40.4) and have participants investigate a single characteristic—such as weight, length, age, or eye color—to assess how representative it might be of the total population of 100 cats.

Hopefully, participants will discuss the idea that *random* sampling is necessary to accurately predict the characteristics of a population.

Collect the Data

How might we select a random sample from the 100 cats database?

Give participants time to offer suggestions. Hopefully, they will arrive at a discussion of the use of coins or dice as tools that might be used to generate random results.

Provide each team of two with a pair of ten-sided dice, each die a different color.

Work in your teams to determine how you can use these ten-sided dice to select a random sample of cats from the database.

Bring participants back together to discuss their thoughts. Guide the discussion so that they suggest using the two digits generated on the dice as the ones and tens places in a two-digit number. Be careful that they don't decide that the *sums* of the dice will work.

For example, suppose participants have a black die and a white die. Letting the black die represent the tens place and the white die represent the ones place will provide all combinations from 00 to 99. (Given their previous work with dice, participants should be able to determine this theoretically.) The results of tosses can be paired with rows in the database. They will have to add 1 to each roll to match the numbering scheme of cats in the database (changing possible results from 00–99 to 01–100).

Explain that in this investigation, we will sample *with replacement* so that a cat may be selected more than once in a sample. This keeps the probabilities the same throughout the selection of the sample, thus mirroring the *What's in the Bag?* activity. ("Teacher Notes" at the right clarifies the with replacement/without replacement distinction.)

You will each roll a pair of dice a certain number of times and record your results. How will you choose what sample size to use?

Using their knowledge from *What's in the Bag?* participants should realize that larger samples demonstrate less variation in terms of matching the population. A sample size of 20 to 30 is fine for this activity.

Each person tosses the dice 20 to 30 times (based on sample size chosen), recording the results for each toss. After adding 1 to each toss, the person marks those cats on Handout 40.1.

Analyze the Data

Our primary question has focused on how well a selected random sample mirrors the population; we haven't asked any questions *about the sample* yet.

We now want to take a closer look at the sample. Think about which attributes you would like to consider—such as body length, tail length, weight, age, and eye color—and then compute statistics and make displays to analyze the data from your samples.

For example, participants may decide to compute the *mean weight* for the cats in their samples. A group display of the mean weights from each sample would help participants to consider the variation within the sampling distribution of mean weights. The actual mean weight for the population can then be computed and compared to the means of the samples.

Teacher Notes

Sampling *with replacement* means keeping all members of the population available throughout selection of the samples. This guarantees that the probability of an individual being selected remains the same throughout the sampling process. In this investigation, the probability of selecting a particular cat is $1/100$ or .01 each time a cat is selected.

Sampling *without replacement* is a problem when the population is small. With 100 cats, when a cat is selected without replacement, it is no longer available as part of the population. After the first draw, the probability for being selected changes from 1/100 to 1/99. If we select a sample of 20 cats, the probability of being selected changes after each draw, becoming 1/89 for the twentieth cat in the sample. Sampling without replacement in a small population changes the probability of choosing the "next one," perhaps noticeably so, and certainly in a way that will affect the estimate of the general population.

When sampling without replacement, we find identical situations with a large population—say 1,000,000 cats—and with a small population. However, the change from 1,000,000 to 999,999 cats is much less significant statistically, or in terms of affecting the probabilities of selection, than is the change from 100 to 99.

Choosing Samples **349**

As another example, participants may decide to compute the *relative frequency for the eye color of green* in their samples. A group display of the percentages of cats with green eyes in each sample would help participants to look at the variation of percentage of green eyes within the samples. The actual percentage for the population can then be computed and compared to the sampling distribution percentage.

Interpret the Results

We want participants to leave this activity with a good sense of what it means to take a random sample and how that might be done. You might even reference the use of a random-number table or the simulation of a random-number generator with a computer as ways to produce samples.

We also want participants to realize that—while there can be variation between a sample and the population—we can attain a relatively high level of confidence using random sampling to predict the behavior of a population. Clearly, sample size has a direct bearing on the reliability of the estimate, as does the notion of random selection and how this method of sampling is implemented.

100 Cats Database

	Name	Gender	Age (yrs.)	Weight (lbs.)	Body (in.)	Tail (in.)	Eye color	Pad color	Breed
1.	Bits	M	0.17	13	20	13	yellow	brown	Maine Coon
2.	Speedy	M	0.17	3.5	15	8	yellow	pink	household
3.	Ginger	F	0.2	2	11	6.5	yellow/green	pink	household
4.	Grey Girl	F	0.2	1.5	13	9.5	gold	pink/black	household
5.	Sasha	F	0.2	2	11	6.5	yellow/green	pink	household
6.	Blue	F	0.25	2	17	8	green	gray	household
7.	Boots	M	0.25	3	15.5	9	brown	black	household
8.	Fire Smoke	F	0.25	2.5	13	7	green/brown	pink	household
9.	Seymour	M	0.25	1.5	13	9	gold	pink/black	household
10.	Simon	M	0.25	2	16	8.5	green/brown	peach/gray	household
11.	Ting	F	0.25	2.5	12	6	green	pink/black	household
12.	Tom	M	0.25	3	14.5	7.5	green	gray	household
13.	Black Foot	M	0.33	1.5	15	7	yellow	gray	household
14.	Booty	M	0.33	1.5	14	7.5	yellow/brown	pink	household
15.	Smoky	M	0.5	1.5	12	7	gray	gray	household
16.	Bradley	M	0.6	11	25	13	yellow	pink/gray	Maine Coon
17.	Nancy Blue	F	0.6	5	14	8.5	blue	gray	Siamese
18.	Bubba	M	0.75	8	19	10	blue/green	pink	Maine Coon
19.	Blacky	F	1	5	18	12	yellow	gray/black	household
20.	Chubbs	M	1	7	22	11	green	pink	household
21.	Duffy	M	1	9	17	10	yellow/green	black	household
22.	Gabriel	M	1	7	14	10	blue	white	household
23.	Pip	M	1	9	17	12	yellow	pink	household
24.	Tabby	M	1	10	20	10.5	green	black	household
25.	Joto	M	1	6.5	14	1.5	gold	pink	Japanese Bobtail
26.	Fuzzy	F	1.25	2	13	8	green	pink	part Persian
27.	Chessis	F	1.5	6	19.5	10	green	brown	Havana
28.	Kahlua	M	1.5	15	13	9	blue	brown	Rag Doll
29.	Kiki	F	1.5	6	21	9	green	black	household
30.	Pink Lady	F	1.5	6.5	15	11	yellow	pink	household
31.	Tabby	F	1.5	7	18	10.5	green	black	household
32.	Koshoshio	M	1.6	10	19	2	green	pink	Japanese Bobtail
33.	Augustus	M	2	10	21	11	yellow/green/blue	pink/black	household

HANDOUT 40.1

	Name	Gender	Age (yrs.)	Weight (lbs.)	Body (in.)	Tail (in.)	Eye color	Pad color	Breed
34.	Chelsea	F	2	9	17	8	yellow	black	household
35.	Jinglebob	M	2	18.5	26	12	blue	pink	Rag Doll
36.	Pepper	M	2	12	17	9	yellow	pink	household
37.	Precious	F	2	12	18.5	10.5	green	pink	household
38.	Feather	M	2.5	13	18	12	green	pink	household
39.	Smokey	F	2.5	8	21	9.5	green	black	household
40.	Boggie	M	3	10	27	10	green	pink	household
41.	Gray Kitty	F	3	9	15	8.5	green	gray	household
42.	Harmony	M	3	12	24	11	yellow/green	black	household
43.	Priscilla	F	3	8.5	23	11	green	pink/black	household
44.	Ralph	M	3	9	23	11	yellow	black	household
45.	Sassy	F	3	8	23	12	yellow/green	gray	household
46.	Samantha	F	3	8	27	12	blue	black	household
47.	Shiver	M	3	12	23	10	yellow/green	pink	household
48.	Sparky	M	3	12	18	8	blue	pink	household
49.	Taint	M	3	11	14	11	green	pink	household
50.	Boo	M	3.5	10.75	19.5	11.5	yellow/green	brown	Somali/mix
51.	Diva	F	3.5	11	20	12	green	pink	household
52.	Hanna	F	3.5	5	12	6	yellow	black	household
53.	Bob	F	4	12	21	13	green	black	household
54.	Cookie	F	4	9	20	11	gold	black	household
55.	Emma	F	4	9.25	18	12	gold	pink	household
56.	Emmie	F	4	7	18	8	green	black	household
57.	Gizmo	M	4	10	23	11	yellow	black	household
58.	Lady Jane	F	4	8.5	19	11	yellow	gray	household
59.	Lucky	M	4	5	19	12	green	pink	household
60.	Prissy	F	4	9	22	10	green	pink	household
61.	Sophie	M	4	12	18	10	green	black	household
62.	Magnum	M	4	8	19	10	green	pink	household
63.	Tigger	F	4	8	17	10	yellow	brown	household
64.	Treasure	F	4	8	19	10	green	pink	household
65.	Amanda	F	4.5	9.75	17	11	blue	gray	household
66.	Matilda	F	4.5	9	19	12	yellow	pink	household
67.	Ethel	F	5	8	19	12	green	black	household
68.	Fluffy	F	5	10	20	12	green	pink	household
69.	Poupon	F	5	16	17.5	10.5	green	pink/black	household
70.	K.C.	M	5	16	24	12	yellow	black	household
71.	Lucy	F	5	10	21	11	green	pink	household

HANDOUT 40.1

	Name	Gender	Age (yrs.)	Weight (lbs.)	Body (in.)	Tail (in.)	Eye color	Pad color	Breed
72.	Oddfuzz	M	5	18	21	9	yellow	pink	household
73.	Peebles	F	5	9	17	11	green	black	household
74.	Tiger	F	5	13	19	11	green	pink	household
75.	Wally	M	5	10	18	12	green	pink/black	household
76.	Ravena	F	6	14	23	12	yellow	pink/black	household
77.	Simon	M	6	13	21	11	green	black	household
78.	Scooter	M	7	16	21	10	gold	black	household
79.	Smudge	M	7	12	21	10.5	green	pink	household
80.	Wiley	F	7	10	16	8	gold	pink/black	household
81.	Gracie	F	8	12	15	12	green	pink	household
82.	Melissa	F	8	11	21	11	yellow	pink	household
83.	Smokey	M	8	10	22	10	green	gray	household
84.	Weary	M	8	15	17	12	green	pink	household
85.	Cricket	F	10	8	22	9	green	black	household
86.	Elizabeth	F	10	9	16	8.5	green	pink	household
87.	Lady	F	10	8.5	17	13	yellow	black	household
88.	Mercedes	F	10	14	22	12	green	pink	household
89.	Midnight	F	10	18	21	9	green	pink	household
90.	Momma Kat	F	10	6	13.75	7.5	chartreuse	gray/white	household
91.	Millie	F	10.5	5	26.5	10	blue	black	Siamese
92.	Charcoal	M	11	12	21	13	yellow	black	household
93.	Miss Moppet	F	11	12	14	12	green	pink/black	household
94.	George	M	12	14.5	21	12	green	black	household
95.	Grey Boy	M	13	12	18	12	yellow	pink	household
96.	Mittens	F	14	10.5	17	10	yellow	pink	household
97.	Peau de Soie	F	15	7	16	13	green	pink	household
98.	Molly	F	15.5	10	17	10	amber	gray	Persian
99.	Strawberry	F	16	14.5	21	10	green	black	household
100.	Alexander	M	18	11	21	11	green	black	household

Includes data from Corwin, R. and S. Friel. *Statistics: Prediction and Sampling.* Palo Alto, California: Dale Seymour Publications, 1990.

HANDOUT 40.2

Total Cat Data Summary

Tail Length (inches)	
Minimum measure	1.5
Lower quartile	9
Median	10.25
Upper quartile	12
Maximum measure	13
Range	11.5
Sum	1004
Mean	10.04

Age (years)	
Minimum measure	0.17
Lower quartile	1.125
Median	3.5
Upper quartile	5.5
Maximum measure	18
Range	17.83
Sum	441.65
Mean	4.417

Weight (pounds)	
Minimum measure	1.5
Lower quartile	6.75
Median	9
Upper quartile	12
Maximum measure	18.5
Range	17
Sum	901.25
Mean	9.013

Body Length (inches)	
Minimum measure	11
Lower quartile	16
Median	18.25
Upper quartile	21
Maximum measure	27
Range	16
Sum	1844.25
Mean	18.443

Female Cat Data Summary

Tail Length (inches)	
Minimum measure	6
Lower quartile	8.5
Median	10.25
Upper quartile	12
Maximum measure	13
Range	7
Sum	544.5
Mean	10.083

Age (years)	
Minimum measure	0.2
Lower quartile	2
Median	4
Upper quartile	8
Maximum measure	16
Range	15.8
Sum	277.7
Mean	5.143

Weight (pounds)	
Minimum measure	1.5
Lower quartile	6
Median	8.5
Upper quartile	10
Maximum measure	18
Range	16
Sum	453.5
Mean	8.398

Body Length (inches)	
Minimum measure	11
Lower quartile	16
Median	18
Upper quartile	21
Maximum measure	27
Range	16
Sum	973.75
Mean	18.032

HANDOUT 40.4

Male Cat Data Summary

Tail Length (inches)	
Minimum measure	1.5
Lower quartile	9
Median	10.25
Upper quartile	12
Maximum measure	13
Range	11.5
Sum	459.5
Mean	9.989

Age (years)	
Minimum measure	0.17
Lower quartile	1
Median	2.75
Upper quartile	5
Maximum measure	18
Range	17.83
Sum	163.95
Mean	3.564

Weight (pounds)	
Minimum measure	1.5
Lower quartile	7
Median	10
Upper quartile	12
Maximum measure	18.5
Range	17
Sum	447.75
Mean	9.734

Body Length (inches)	
Minimum measure	12
Lower quartile	16
Median	19
Upper quartile	21
Maximum measure	27
Range	15
Sum	870.5
Mean	18.924

INVESTIGATION 41

How Tall Are You?

Overview

Participants collect data about their height and examine the data in several ways, including using a *univariate analysis,* or one-variable analysis. The study of how data collected about two variables—estimated height and measured height—are related in a *bivariate data analysis.* The bivariate analysis will include displaying the data in a scatter plot.

Assumptions

Participants are familiar with interval (ratio scale) data and with a variety of representations for displaying and summarizing such data (for example, stem plots, histograms, and box plots).

Goals

Participants explore the concepts of univariate data and bivariate data. In particular, they

- identify and describe patterns of variation in data

- recognize linear trends in bivariate data

- recognize various degrees of linear association

- examine the mathematical relationships $y = x$, $y > x$, and $y < x$ in context

- use linear interpolation to make predictions

Reference

Perry, M., and G. Kader. *STAT-MAPS* (NSF: MDR-9150117), North Carolina: Appalachian State University, Forthcoming.

Materials

Meter sticks or metric measuring tapes, oversized graph paper with a 1" × 1" grid, ½" stick-on dots (in 2 colors), calculators

Fun Facts

The tallest man in medical history is Robert Pershing Wadlow (Alton, Illinois), measuring 8 feet 11.1 inches. The tallest woman in medical history is Zeng Jinlian (Hunan Province, China), measuring 8 feet 1.75 inches.

The shortest man in medical history is Gul Mohammed (Delhi, India), measuring 22.5 inches. The shortest woman in medical history is Pauline Musters (Ossendrecht, Netherlands), measuring 19 inches.

Mathews, P., M. D. McCarthy, M. Young, and N. D. McWhirter. *The Guinness Book of Records 1993*. New York: Bantam Books, 1993, pp. 143–45.

Teacher Notes

Here is one way to make measuring move quickly. Tape three or four meter sticks to the wall vertically, positioned exactly 1 meter above the floor. A person measures his or her height against the wall and adds 100 centimeters to his or her *measured* height.

Developing the Activity

Participants will first estimate their height, then measure their actual height and make comparisons of the pooled data.

Pose the Question

How tall are you? How tall is a typical person in this group?

If there are a sufficient number of each sex—eight or more—in the group, you may want participants to compare data between males and females.

How do you think the heights of males and females in this group compare?

Have each participant state her or his estimated height, in inches, to the nearest half inch. Allow them to discuss your questions.

When stating your height, how accurate do you think you were?

Do you think males and females differ in their accuracy when stating their heights? (This question is optional.)

Collect the Data

Ask participants to convert their stated heights to centimeters by multiplying by 2.54 and rounding to the nearest centimeter. Then have them work in pairs, measuring their partner's height to the nearest centimeter.

Analyze the Data

Univariate Analysis (Measured Height)

Help participants to summarize their data using one-variable analysis.

What are the largest and smallest measured heights? What are some appropriate ways to represent these data? What scale would you use for your graphs?

Have participants make at least two appropriate displays of their data. Scaling both the horizontal and vertical axes in increments of 1 or 2 should produce a good graphical display.

What is a typical height for someone from this group? Explain your reasoning.

Comparative Analysis (Measured Height)

If you chose to have participants collect data about males versus females, have them perform a comparative analysis by analyzing differences between appropriate numerical summaries and graphical representations of the data.

Bivariate Analysis (Stated Height versus Measured Height)

Now lead participants through a bivariate analysis.

What are the largest and smallest stated and measured heights? Let's make a scatter plot of the data.

As a group, make a scatter plot on oversized graph paper. Let the horizontal axis be measured height (in centimeters) and the vertical axis be estimated height (in centimeters). (In terms of identifying the degree of linearity in the data, it doesn't really matter which variable is x and which variable is y. However, it does make the interpretations discussed here easier when examining where a point is relative to the line $y = x$.)

What scale should we use for the axes?

Since the actual and stated heights are similar, the same scale can be used for both axes.

Have each participant place a stick-on dot at the location that marks his or her (measured, stated) height. If you are also investigating males and females, use dots of two different colors. Here is how the scatter plot might look.

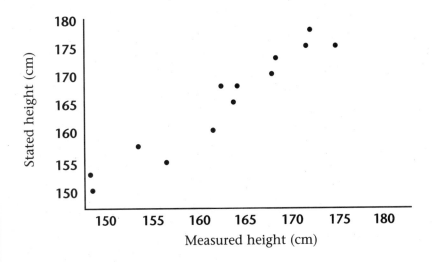

Teacher Notes

Since the points are color-coded according to sex, you can identify the sex of anyone who is over- or understating his or her height. Some studies have shown that men often overstate their height.

How Tall Are You? **359**

Interpret the Results

Univariate Analysis

What is a typical height for someone in this group? How does your height compare with this typical height? How much do the heights from the entire group differ from this typical value? Is there a pattern in the way the heights differ?

Measured height will typically fall between 150 and 200 centimeters.

There are several graphical displays that are appropriate for this analysis, including stem plots, histograms, and box plots.

You might want to have participants experiment with scaling. For a stem plot, use intervals of 5 or 10. For a histogram, intervals of other sizes are possible. If participants construct a stem plot, discuss the shape of the data displayed in the stem plot.

Unless there are some extremely tall or short heights, the mean or median should be appropriate for representing a typical height. If there are outliers, the median may better represent what is typical. You might compare the values of the mean and median.

Are the mean and the median similar? Explain why or why not.

Comparative Analyses (optional)

How does a typical male height compare with a typical female height? How does the pattern in male heights compare with the pattern in female heights? How does the amount of variation in the heights compare between the two groups?

Males are typically taller than females; however, there is probably a range of heights shared by both males and females. This range can be identified by comparing across two or more graphical displays. The box plot is especially useful for comparing not only the median but also the ranges for the different quartiles of the data.

Bivariate Analysis

How do the estimated and measured heights compare?

If people stated their heights accurately, their measured height would equal their stated height. Mathematically, this is the relationship: Stated Height = Measured Height, or $y = x$.

Graph the line $y = x$ on your scatter plot. Examine the data points around this line. How close are the points to the line? Identify any points that are far from the line.

Sample Questions

Questions about the shape of the data:

Are there any clusters in the data? Any peaks?

Do any values "stand apart" from the main body of data?

Is there symmetry in the data?

The data in the scatter plot should show a fairly strong linear trend, clustering close to the line Stated Height = Measured Height or $y = x$.

What does it mean when a point is on or near the line? What does it mean when a point is far above the line? What does it mean when a point is far below the line?

Points on or near the line represent people who accurately stated their height ($y = x$). Points far above the line correspond to people who overstated their height ($y > x$). Points far below the line correspond to people who understated their height ($y < x$).

INVESTIGATION 42

Are You a Square?

Overview

Participants collect data on their arm span and examine the data in several ways. The notion that the size of one body part is related to the size of another is examined.

Assumptions

Participants are familiar with interval data and with a variety of representations for displaying and summarizing such data (for example, stem plots, histograms, and box plots).

Goals

Participants explore the concepts of univariate and bivariate data. In particular, they

- identify and describe patterns of variation in data

- recognize linear trends in bivariate data

- recognize various degrees of linear association

- examine the mathematical relationships $y = x$, $y > x$, and $y < x$ in context

- use linear interpolation to make predictions

Reference

Perry, M., and G. Kader. *STAT-MAPS* (NSF: MDR-9150117), North Carolina: Appalachian State University, Forthcoming.

Materials

Meter sticks or metric measuring tapes, oversized graph paper with a 1" × 1" grid, ½" stick-on dots (in 2 colors), height data from Investigation 41, calculators

Transparency 42.a

Variations

This activity might be especially interesting when performed with "special" groups such as a basketball team. The data from the special group can be compared with that of another group using color-coded points.

Another variation is to use color-coded points to compare the data between older and younger people.

Developing the Activity

Participants will measure their arm spans and, using their group's height data from Investigation 41, *How Tall Are You?* make comparisons of the data.

Pose the Question

Display Transparency 42.a.

Leonardo da Vinci's drawing of the human physique scales the human body in a way that proportionately encloses it within a square. This suggests that a person's arm span is the same as her or his height. How close to being a "square" do you think the people in this room are?

Collect the Data

Have participants work in pairs, measuring their partner's arm span (fingertip to fingertip) to the nearest centimeter.

Analyze the Data

What are the largest and smallest measured heights? What are the largest and smallest measured arm spans?

Have participants make a scatter plot of the data, with height on the vertical axis and arm span on the horizontal axis. Scaling both the horizontal and vertical axes in increments of 1 or 2 should produce a good graphical display. (In terms of identifying the degree of linearity in the distribution of the data, it does not matter which variable is x and which variable is y. However, this labeling does make the interpretations easier to understand and locate when examining where a point lies relative to the line $y = x$.)

What scale will you use for the axes?

If participants have collected male versus female data, have them use one color for male data points and another for female data points.

Interpret the Results

How do height and arm span compare?

Measured height and arm span will both typically fall between 150 and 200 centimeters.

364 *Are You a Square?*

Explain that if someone's height and arm span are equal, we could call that person a "square." Mathematically, this is the relationship: Height = Arm Span or $y = x$.

Graph the line $y = x$ on your scatter plot.

Examine the data points on and around this line. How close are the data points to the line? Identify any points that are far from the line.

The data in the scatter plot should show a fairly strong linear trend, with the data clustering close to the line Height = Arm Span ($y = x$).

What does it mean when a point is on or near the line? What does it mean when a point is far above the line? What does it mean when a point is far below the line?

Points on or near the line indicate people who are nearly square ($y = x$). Points far above the line correspond to people who are significantly taller than their arm span ($y > x$); these people might be described as "tall rectangles." Points far below the line correspond to people who are significantly shorter than their arm span ($y < x$); these people might be described as "short rectangles."

If participants have investigated male versus female data, have them compare those data.

Does one sex appear to be more likely to be square than the other? Explain your observations.

You might compare the strength of the linear trend in the data with the stated versus measured height data from Investigation 41, *How Tall Are You?*

Does the distribution of the points in one plot appear to be more linear than those in the other?

Leonardo's Square Man

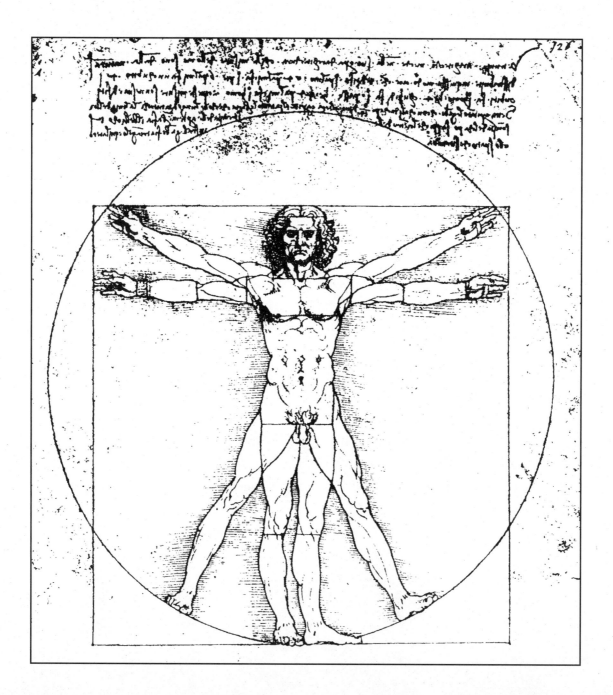

From Leonardo da Vinci's *The Golden Notebook*

INVESTIGATION 43

From Footprint to Stature

Materials

Meter sticks or metric measuring tapes, oversized graph paper with a 1" × 1" grid, ½" sticker dots (in 2 colors), height data from Investigation 41, calculators

Overview

Participants collect data on the length of their feet and examine the data in several ways. They again encounter the idea that the size of one body part may be related to the size of another.

Assumptions

Participants are familiar with interval data and with a variety of representations for displaying and summarizing such data (for example, stem plots, histograms, and box plots).

Goals

Participants explore the concepts of univariate and bivariate data. In particular, they

- identify and describe patterns of variation in data

- recognize linear trends in bivariate data

- recognize various degrees of linear association

- examine the mathematical relationships $y = x$, $y > x$, and $y < x$ in context

- use linear interpolation to make predictions

Reference

Perry, M., and G. Kader. *STAT-MAPS* (NSF: MDR-9150117), North Carolina: Appalachian State University, Forthcoming.

Developing the Activity

Participants will measure their foot lengths and, using their height data from Investigation 41, *How Tall Are You?* make comparisons of the data.

Pose the Question

Forensic anthropology is the study of the characteristics and customs of human beings based on evidence found in human remains.

If the length of a person's foot is known, for example, what other physical characteristics do you think can be determined? Specifically, can a person's height be estimated from that person's foot length?

Collect the Data

Have participants work in pairs, each measuring the length of his or her right foot (with socks on) to the nearest millimeter.

Analyze the Data

What are the largest and smallest measured heights? What are the largest and smallest measured foot lengths? How do they compare?

Measured height will typically fall between 150 and 200 centimeters, and measured foot length between 200 and 300 millimeters.

Have participants make a scatter plot of the data, with height on the vertical axis and foot length on the horizontal axis.

What scale will you use for the axes?

A horizontal scale in increments of 4 and a vertical axis in increments of 2 should produce a good graphical display. If participants have collected male versus female data, have them use one color for male data points and another for female data points.

Interpret the Results

How are height and foot length related?

Examine the data points in the scatter plot. How linear do the data appear to be? Sketch a line that goes roughly through the approximate center of the data as it is distributed across the scatter plot.

From the line you drew, predict your height based on the length of your foot. How close is the predicted height to your actual height?

Now predict the height of a person with a foot length of 250 millimeters.

Bivariate Data Analysis (optional)

In this analysis, participants investigate how the sizes of certain body parts compare between males and females, and how the sizes of the various body parts are related.

Working with a partner, measure the circumference of (the distance around) your

- *head*
- *neck*
- *wrist*
- *waist*
- *ankle*

Once participants have collected their data, have them conduct bivariate analyses. If you have a large enough sample—a minimum of eight males and eight females—participants can compare data between males and females.

Pick two variables (measurements that you made), and examine the relationship between them.

INVESTIGATION 44

Cats Revisited

Overview

Participants revisit the cat data from Investigation 40, *Choosing Samples*, pairing variables such as body length and weight, tail length and weight, and age and tail length to look for relationships between them. They will discuss that relationships among measures of cat length, weight, and age do not emerge as obviously as relationships in the data participants collected about their own height, arm span, and foot length. The bivariate data analysis includes displaying the data in scatter plots, which, when appropriate, allow us to make predictions for future observations.

Assumptions

Participants are familiar with interval data and with a variety of representations for displaying and summarizing such data (for example, stem plots, histograms, and box plots).

Goals

Participants explore the concepts of univariate and bivariate data. In particular, they

- identify and describe patterns of variation in data

- recognize linear trends in bivariate data

- recognize various degrees of linear association

- examine the mathematical relationships $y = x$, $y > x$, and $y < x$ in context

- use linear interpolation to make predictions

Reference

Perry, M., and G. Kader. *STAT-MAPS* (NSF: MDR-9150117), North Carolina: Appalachian State University.

Materials

Cat data from Investigation 40, oversized graph paper (with a 1" × 1" grid), ½" stick-on dots (in 2 colors), computers and graphing software (optional), calculators

Developing the Activity

Participants have done quite a bit of work with cat data in earlier activities. Here they explore relationships among cat length, weight, and age. It is particularly helpful if the cat data can be analyzed using a computer graphing program, as then graphs won't need to be made by hand.

Pose the Question

The data set is the *100 Cats Database* (Handout 40.1). Work with participants to generate questions about how certain variables might be related. Participants may want to first investigate data from all cats, followed by just male cats, and then just female cats.

Collect and Analyze the Data

Have participants work in teams of two to explore one of the questions by graphing the data. If computer graphing software is available, participants may explore questions as a group.

Interpret the Results

One way to think about whether two variables are related is to ask, "If I know values for one variable, does this help me estimate the paired values of the second variable?" If the answer is yes, then we say that the two variables are related to each other: A change in one is paired with a change in the other in some patterned way. We want to describe the *strength* of the relationship. Is the pattern of change very clear? Do the points on a scatter plot cluster closely around some visualized line or are they more spread out with less of a focus around a visualized line? These guidelines can be used with the cat data.

Sample Questions

Is a cat's body length related to its tail length?

Is a cat's weight related to its age?

Is a cat's body length related to its weight?